EVOLUTION

Science or Ideology?

EVOLUTION
Science or Ideology?

İrfan Yılmaz

Translated by Neva Tuna

TUGHRA
BOOKS

New Jersey

Originally published in Turkish as *110 Soruda Yaratılış ve Evrim Tartışması* in 2008

27 26 25 24   2 3 4 5

Published by Tughra Books
335 Clifton Avenue
Clifton, New Jersey, 07011, USA

www.tughrabooks.com

Library of Congress Cataloging-in-Publication Data is available

Hardcover ISBN: 978-1-59784-118-4

Paperback ISBN: 978-1-59784-966-1

TABLE OF CONTENTS

PREFACE

O ver the past thirty-seven years of my career as an academic in biology, I have encountered countless questions
on evolution and creation. In this book I have endeavored
to answer these questions. As I was raised with the belief that the
whole human race is from Adam and Eve and that neither human
beings nor other species have ever been exposed to any evolution, I
remember being shocked to see the sketches of how apes gradually
turn into human beings. The skulls found proved this fact with certainty! Our teachers told us that evolution is no longer a hypothesis
but a scientifically proven fact. However, none of my teachers who
spoke in defense of evolution could influence me during my elementary and high school years. When I started to study Zoology and
Botany at the university I realized that the worst was yet to come.
There was the compulsory Evolution Course given at the department.
What is more, the courses on Systematic Zoology and Botany,
Comparative Anatomy, Physiology, Histology, Embryology and
Genetics were all taught in alignment with evolution scenarios. The
"idea of evolution" was enforced to such an extent that it was converted into a worldview, ideology, dogma – and even a religion to be
believed in. Making matters still worse, there was a shortage of publications presenting differing or opposing ideas. Nor could I find
satisfying information in religious texts to deal with the issue and
which does not ignore the biological side of the subject or scientific
progress of the time.

With the bombardments I was exposed to during my college years, I was seriously at risk of perdition; that is, I came closer to dying as an atheist. Thanks to a friend of mine I narrowly escaped such a death because of the book he gave me: Nursi's *Nature: Cause or Effect?* was the first work that changed the flow of my life with its convincing and strong language on the existence of God and His creation of everything with His infinite knowledge and willpower. Later, in 1976, I got hold of *İlmi Gerçekler Işığında Darwinizm* (Darwinism in the Light of Scientific Truths). This book was a translated compilation of John N. Moore's "On Chromosomes, Mutations, and Phylogeny"[1] and A. N. Field's *The Evolution Hoax Exposed*.[2] This was indeed the first book I had read that dealt directly with evolution.

In the following years, I started to follow the publications in favor of or opposed to evolution, especially those published in the US. As time has elapsed, I have seen an increasing number of publications with arguments for or against evolution. After I became an academician, I readily volunteered to teach the Evolution Course when all the other faculty members hung back. Other professors ducked out of teaching this course because they were shying away from arguing with students and lacking assertiveness and knowledge, they could not face up to differing views and controversies. But I pressed to teach the course as I had focused on this controversial issue for years. I wanted to teach it not in a strictly imposing way that shows evolution to be "an absolutely proven law" but in an objective way with a democratic approach. I also wanted this course not to be given under the title of "Evolution" but under such titles as Philosophy of Biology or Biophilosophy, given that the subject is not predicated on empirical research or verified through experiments and observation.

Those who closely follow scientific developments know that thought and movements opposing evolution have been growing rap-

---

[1] Moore, "On Chromosomes, Mutations, and Phylogeny," *Creation Research Society Quarterly*, December 1972, pp. 159-171.
[2] Field, *The Evolution Hoax Exposed*, TAN Books & Publishers, 1971.

idly, especially in recent decades. In many countries, especially in the US, scientists have started to increase their vocal opposition to the idea of evolution—a concept which has, in fact, gained ideological peculiarity rather than biological reality over the years—and are publishing numerous articles against the theory of evolution through various institutions and foundations. The effects of this process of transformation all over the world have finally reached Turkey, where evolution has long been converted into a dogma with a materialistic and positivist mentality—a dogma which has been taught and imposed most adamantly. Against the evolutionists' recent collection of signatures in order for evolution to be taught unilaterally with no reference to creation at all, I feel that it is necessary for me to speak up.

Throughout my professional career, I have pointed out all the claims, evidence, and argumentation for and against evolution in my lectures. I have let my students express their views freely without intimidating them with low grades or imposing any pressure on them. I have seen that the lessons were more efficient and interactive and the students had so many questions to ask and really wanted to deal with the issue. Yet in spite of all this, I became the butt of the preponderant materialist and positivist mentality in my country. Against the aggressive attacks of those who wished to take the issue of evolution out of the scope of science, fixing it entirely within an ideological framework and making it a "scientific camouflage" for their Marxist and atheist mentality, I have continued to write articles for various publications under different pen names. I am also planning to write my memoirs of the difficult and painful experiences I have gone through as a scientist of zoology. I have never given up striving for this cause, even though I was treated unjustly and prevented from being a full professor for nine years.

I shall never forget the moment when my speech was interrupted and I was forcefully removed from the lectern at the academic conference on "The Problems of Biology Education" held by the Faculty of Sciences at Istanbul University. Not only was I prevented from presenting my paper, I was also punished by the omission of my paper from the conference proceedings book. Another case of such an uncivilized approach victimizing persons who disagree or choose to talk about something different was that of Dr. Adem Tatli, the professor of botany, who was dismissed from his position at the university just because he argued against and wrote about "the evolution taboo."

Even though they tried really hard to suppress my arguments and intimidate me into resigning or to dismiss me from teaching, they could not find any student to attest to the false claim that I could not teach my courses objectively. To this end, they even used a couple of atheist students to record my lectures secretly, but all the ambushes they planned to incriminate me have come to naught. Nowhere in the world has a scientist been exposed to academic persecution to such a degree, including in the communist former Soviet Union. As the old adage says, "The truth will out," so this oppression and persecution cannot last and people will finally express their ideas freely.

Doesn't all this show how ideological the idea of evolution has become? For this reason, I have given my work the subtitle, "Science or Ideology?" This book will not be able to put an end to this struggle—and, as a matter of fact, it should not be expected to do so. However, I cannot remain silent in the face of unrelenting attempts to place an atheist ideology at the foundation of the education system in the name of "science" by pressure, manipulation, fabrication and fear. It is to be expected that many scientists who have long been suppressed in an antidemocratic atmosphere in the country and especially by the directors of the Turkish Higher Education Council, who maintained and enforced persistent pressure and control through the use of harsh measures between 1994 and 2008, will be able to express their ideas freely. The articles and books they will produce will, hopefully, be more in line with the democratic ambiance of a country preparing to enter the EU.

Since my years as a teaching assistant I have become more aware of an important fact: The articles and books published on evolution did not only comprise those written in defense of evolution. On the contrary, there were many publications against evolutionary theory. However, the vast majority of university students were not even aware of them since publishing was mostly the monopoly of certain interest groups tightly organized, and advocating certain thoughts in disregard of other views. What is more, there was no internet at the time. I lived through the times when university students were silenced because they queried with their limited knowledge the dictatorial teaching of evolution and were reviled by the professors as reactionary bigots.

Over years of discussion with local and foreign academicians, I have come to realize that academicians have differing views of evolutionist thought. Whether they are atheist, Muslim, Christian or agnostic, academicians often do not mean exactly the same thing when they talk of evolution. Though it seems that there is dichotomy between theistic and atheistic attitudes towards evolution, in reality there exists a range of views about evolution. Accordingly, different people advocate and adhere to different versions of evolution in accordance with their own worldview, faith and philosophy.

Those academicians that have a strong faith in God and devotion to their religion, whether they are Muslim, Christian or Jew, believe that evolution is not a scientific theory but a dogma which is being used as an instrument to deny God, and which has been converted into a worldwide perception and a belief system.

There are also those who have little devotion to their religion and are not sensitive about how evolution affects their faith. Though they have faith in God, they accept evolution as a scientific theory and believe that God has created existence in accordance with the mechanisms and principles of evolution.

For atheists, on the other hand, evolution is undeniably clear and is an absolutely proven fact. For such people, evolution is the indispensable foundation of everything and the necessary condition shaping the whole world. As for the agnostics, though they believe in the ongoing process of evolution, they remark that it is impossible to say anything certain about the beginning of the universe or life.

There are also remarkable differences of opinion among the believers in heavenly revealed religions. Though they all have faith in God and stand against the idea of evolution, it is important, at this point, to stress Islam's unique approach to the elucidation of creation, satisfying both the mind and the heart. I have tried to be fair and conscientious in my evaluations, always keeping in mind the possibility that my Muslim perspective may cast doubts in the mind of some people who may question my objectivity. Considering the tolerant approach of the papacy toward evolution and their emphasis that evolution can indeed be reconciled with the Christian faith, I am convinced that the Islamic faith in God and the belief in the Islamic faith in the intricate and infinite manifestations of such Divine Attributes and Names in the whole of existence as the All-Knowing, the All-Powerful, the All-Originating, the All-Governing and the All-Omnipotent explain the phenomenon of creation best. In addition, belief in God as a sort of static entity, the description of the universe as a clock which has been wound, or the belief that God "leaves the universe to the coincidental moves of evolution" are all far from the Islamic faith. Apart from those who believe in God and oppose evolutionary theory, those who are too concerned with being "scientific" and who pull back from their faith for the sake of sacrosanct "science" constitute the majority among the followers of other heavenly revealed religions.

It is also to be noted that there are some Muslim scholars who argue that "evolution" can be elucidated with Islam and even claim that some Qur'anic verses allude to it. However, they are deprived of supporters as the related verses refer to creation and spiritual progress as a matter of fact. After this preamble, let us focus on the issue now.

ENDLESS QUESTIONS

The questions, "Where did we come from?", "How did we come to this world?", and "Where are we going?" are simply the most deeply pondered questions of those who think. Other than the knowledge about creation delivered by the heavenly revealed religions, the hypothesis of "evolution" is probably the most common claim offered in answer to the first question.

Such questions are the hallmark of a thinker, and what raises them is human curiosity. In fact, the drive behind all invention and discovery is the devotion to research and investigation which stems from this curiosity. We observe the world and the universe that we live in, and we gather information about things with this curiosity; then we analyze this data using our intellect and logic. Some of that information may not be very important for us—that is, it might not affect our lives either positively or negatively—but it still might be a crucial piece of information for those who specialize in and limit themselves to that particular field of study. For example, for a food engineer, how radio waves are emitted or how a satellite antenna works does not hold any significance for his or her profession. Similarly, an electronics engineer does not usually wonder about how and which toxins certain bacteria secrete; however, if she or he experiences food poisoning, then they will go to a doctor and may learn about which bacteria caused the poisoning.

On the other hand, the questions which interest all human beings—in that they are related to the reason behind our being in this world, how we became living beings, and what will happen to us in the future—will always remain important to us. We hope that by getting sensible answers to them at different times and under varying conditions, we will be fully satisfied—including in our minds and hearts—after we secure their strong approval in our conscience. Our logic and intellect combined with our souls, demand that we ask such questions and seek reasonable answers to them in order to obtain peace in our inner selves.

Uneducated people might not be curious about these questions, and they may be satisfied with what can be learned from parents or grandparents. Such people can find peace and comfort in ordinary life in proportion to their personal beliefs. They will not have any doubt about their religion and will simply find comfort in the certainty that God creates everything in a way that He wishes, and He extinguishes everything according to His will. However, the possibility of running across such a person these days is very small since new communication devices have changed the world into a huge community, and educational developments have brought all kinds of scientific discussions even to very small villages far away from urban centers. Now, all types of information, right or wrong, reach people via many kinds of media devices. While some questions are being answered in this media bombardment, people have also become confused by incorrect information and prejudicial comments about their most important values and perspectives, and thus their central way of thinking is being turned upside down.

Most people's minds have been confused and the basis of their beliefs has been shaken by this mass media deluge affecting the whole world. The belief that "religion and science contradict" has been reinforced through misleading propaganda, and the dominant idea that all creatures, including humans, came into being by themselves or by the random influences of forces, called "causes," and evolved has led the world into a terrible downfall. In the struggle between theism and atheism, a struggle which started at the beginning of human life on Earth, the very important tools of science and technology have been used to support atheism, following the lead of dominant materialistic philosophies, and with the assistance of the mass media. The basic arguments of atheism—namely, the expressions of materialism, coincidence and nature—have never varied since the Ancient Greek times. Only now, the representation of atheism has been polished with science; hence people have become misguided, and generations have been thrown into an emptiness of belief and faith. As a result, humanity is drifting towards catastrophe, falling ever deeper into a crisis of belief.

Being only a biological hypothesis, "the idea of evolution" has been converted into a belief system or worldview, and global communities have been shaken by the directive to believe in it. In this book, the degree to which the idea of evolution is, or is not scientific, the truths and deceits of which it is comprised, the artifice which has been incorporated in it, and subjective comments which have been made about it will be explained one by one. Although the topics are dealt with briefly and can be understood by anyone having a basic knowledge of high-school biology and general science, I plan to expand at a later date on the issues which readers may bring to my attention for further clarification.

In this work the emphasis is primarily on the main ideas, rather than on technical details. The issues can be analyzed in a much shorter way as judgmental sentences. However, in that case, it might seem to have been written with an ideological "obsession," in much the same way as evolutionists write their ideas without demonstrating the truth of their assertions; so I have preferred to explain some of the topics in detail.

I would like to thank the team that has expended considerable effort on this book. My sincere thanks go especially to geologist Dr. Ömer Said Gönüllü for his help updating the astronomical and geological information. I am also very grateful to all the employees of Tughra Books for their meticulous work on the translation, editing, cover, design and publication of this book.

İrfan Yılmaz
September 24, 2008

1

A Perspective on Science and Belief

A PERSPECTIVE ON SCIENCE
AND BELIEF

J ust as believing in creation necessitates believing in the Creator, Who has infinite knowledge and power, sufficient to do everything, and Who is eternal and everlasting, believing in evolution conversely necessitates disbelieving in the Creator, the One God, and putting unconscious and senseless laws of nature in His place. In this case, atoms and coincidence will be considered to have intellect, sense and knowledge, and those factors are placed in the position of deity. Many people who do not have the true knowledge of God's names and attributes claim that they believe in God; however, in thinking that the Theory of Evolution does not contradict the belief in One God, they become disobedient or atheistic without even realizing it. For, even some of those who accept that God created the universe in the beginning have a belief that after the first creation, God let the universe run by itself, like setting up a clock and that He did not intervene afterwards—that He simply assigned everything to the laws of nature, and that those laws can make all the creatures, plants, animals, and even human beings, come to life coincidentally, all by themselves.

At first glance, many people might not realize that the Theory of Evolution causes disobedience to God. To be able to achieve this, the idea of evolution has been erected as an elaborate montage, decorated very well and hidden under an artificial compilation of logic. However, when you dig out the subject by questioning it step by step, in the end, you realize that the foundation of the Theory of

Evolution is chance. You might become confused when you first notice that chance is the basis of an idea which has dominated the scientific world for 150 years. Even though it does not make sense to you how millions of living beings—and their biological systems, organs, tissues and cells—came into being through the unconscious forces of nature flowing like a stream, and through the coincidental reactions of chemical elements, it is expected that you will be helpless in the face of "science," which has effectively been transformed into a taboo, and the propaganda of the mass media, which presents evolution as an event proven with certainty.

This represents such tragic, scientific and media exposure that it is hard to avoid becoming a disbeliever without having a very strong knowledge of God. In order to oppose any kind of distortion or lie under the pretext of science, all knowledge and information that is produced against evolution is derided as being "old-fashioned," "unscientific," "unprogressive," and "dogmatic." If you are a scientist yourself, worse things may happen to you, for the establishment can do anything to obstruct your academic career; you may face a media lynching, and any kind of deceit may be used to dismiss you—simply because you attempt to question the dogma and you have tried to come up with an alternative idea. Your opponents will also argue that what you have written in this field does not have any value since you believe in God. According to this distorted idea, "a scientist cannot believe in God," "evolution is a certain phenomenon which cannot be questioned," and hence "only after you accept that there is evolution can you discuss how it happens."

These explanations should not be considered to be exaggerated; in fact, the present author has personally experienced such unfortunate incidents in the most painful way.

Before starting the debate about "creation versus evolution," and prior to discussing the scientific evidence for evolution, a basic matter has to be clarified. If those who defend the Theory of Evolution had approached this idea as a matter of belief, no one could say anything to them, since belief and faith cannot be disputed. No matter what a person believes in, she or he should be respected. As those who believe in God and creation have a right to believe and have trust in God, those who believe in the Theory of Evolution and creatures rising to become living beings have the right to believe in the evolution and existence of creatures through natural forces. Some people might be atheistic, agnostic or theistic— and this is a matter of belief which concerns only those people. On the other hand, they do not have a right to force people to accept what they believe as "proven, absolute facts," or as "scientific determinations against which there cannot be any contrary opinion"— nor, under the guise of "science," to label those who believe in God as being "unprogressive."

Nowadays, no one disputes the existence of gravity, air pressure or the expansion of metals; rather, many physical events are explained by these phenomena, which are expressed by formulas, and problems are solved using such formulas. Being the subject of science, we all know that these matters have nothing to do with one's belief. However, the existence of angels and the jinn, for example, is not a scientific subject; rather, it is a matter of faith. Such concepts are generally not studied with the methods of science, which are valid in a limited field, and they are not observed or experienced objectively—rather, they are personal experiences gained by using one's heart and intuitive faculties, and thus related to a person's belief.

The idea of evolution is neither similar to the laws of physics mentioned above, nor does it have the character of being experienced by a person's heart and soul. It is simply the subject matter of belief, attained solely by observing the abundance of creatures in nature and by interpreting some of the changes in living beings.

From this point, the "Hypothesis of Evolution" is just like a matter of belief—and yet, it is also a kind of creed. Religion is the main source for the values which shape a person's life. If a person believes in God, others will observe the reflections of faith in every moment of that person's life. So, too, belief in Evolution influences the lives of those who adhere to it, and it is a main cause, just like a religion, in shaping their lives. And certainly, those who believe in Evolution should have the freedom to "practice" and teach their beliefs. But it cannot be acceptable for them to attack those who believe in divine religions and do not think in the same way they do, or for them to consider their opponents to be enemies of science.

Historically, biology had been descriptive; namely, it had been seeking to explain what was present since past times. It had tried to reach deeper and arrive at universal knowledge by observing the excellent design and harmonious art in living beings, and by gathering information about the structures and operations of the systems, organs, tissues and cells of plants and animals that we observe in nature. Further, by analyzing this information, it had been trying to understand general principles at higher levels. The beauty of a living being that was examined, the perfect structures that are free from any imperfections, and the holistic order and operation of the ecosystem, used to urge every logical and brilliant person to search for a Creator.

Contrary to all this, the agreement between religion and science was broken by the idea of evolution's assignment of these perfect structures and mechanisms to mindless and unconscious random operations of natural laws, instead of assigning the creation of them to the Creator. Scientific thought became detached from religious sanctity; simultaneously, science became a taboo which could not be challenged. The practice of using biology's interpretation of the beauties of life to cause faith in people's hearts came to be degraded to viewing life as a phenomenon that arose by itself. Serving as a cause for technological improvements in astronomy, engineering and medicine, constructed knowledge in physics and chemistry increased the courage of those who gave a sense of holiness to science and caused religious people to become timid and develop the urge to refrain from science. However, these developments were the fruits of the talents that God gave to humans—of hard work, effort and devotion to research and experimentation.

God created the human being as the most perfect of all creatures and gave them the authority to manage things on Earth, by raising them to the position of being a "caliph" over them. Meanwhile, humans began developing new technologies for their own happiness and comfort, using the knowledge that God gave them, but they claimed these improvements as their own successes, and they attributed every event to the laws of nature while rejecting the Creator.

Christianity itself weakened under the common pressure of all these factors; it was not able to recover the authority it lost with the Renaissance and Reformation, and it fell to materialistic and positivist ideas of the "new science perception." In this atmosphere, the "Hypothesis of Evolution" was made into a primary focus with the assistance of the mass media and other "dark forces"—hidden agendas and influences—which controlled the scientific world. Thus, every discovery, and each type of information or data obtained, came to be interpreted from an evolutionary perspective—and every scenario and piece of fiction that was written was commented on in a way that supported evolution until it became the dominant paradigm. In this way, the idea of evolution, which put on the apparel of being scientific, was given the most prominent places in science books. As Rifkin aptly mentioned, evolutionary theory has been enshrined as the centerpiece of our educational system, and elaborate walls have been erected around it to protect it from unnecessary abuse. Great care is taken to ensure that it is not damaged, for even the smallest rupture could seriously call into question the entire intellectual foundation of the modern worldview.[1]

For his part, Huxley spoke nonsense with confidence when he stated that the Darwinian Theory of Evolution is no longer a theory but a fact. For him, no serious scientist would deny the fact that evolution has occurred just as he would not deny that the Earth goes around the Sun.[2] Yet it is very strange that though it claims to be "scientific," evolutionary theory has never respected a very essential criterion for scientific studies: to listen to and try to understand

counter-arguments. In addition, evolutionists strove to make inef-
fective any possible resistance to their ideas by excluding counter-
arguments as "unscientific" or even "fanatical"; thus, evolutionary
theory earned a "sacred" immunity in time. With the knowledge
God blessed them with humankind has developed technology for
pleasure and comfort; nevertheless, they have attributed all the
progress to their self-achievement and to laws in nature, and they
have rejected belief in the Almighty Creator.

Darwin was an agnostic as to some of his attitudes, and he was
a deist as to other aspects, but he was actually a faithful Christian
before he mentioned the hypothesis of evolution, and he even went
to ministry school. Yet, evolutionary thought had a huge impact in
the scientific world after it assumed its shape and was published as a
book. Basically, a couple of major factors can be mentioned regarding the acceptance of evolutionary thought in Europe and about the
rapid spread of the theory among the scientific community.

First of all, the starting point for Darwin was the phenomena
that he observed in nature. At the beginning of the long voyage that
he took on the ship called *Beagle*, Darwin was amazed by the diversity of living beings, the richness of variations in species, and the
perfection of various adaptations in subspecies. However, the
weakness of his religious understanding, namely his lack of knowledge of God's names and attributes—knowledge that is particularly
exclusive to Islam—left him incapable of properly appreciating or
interpreting this prosperous nature. Meanwhile, Darwin had a
fairly limited realization of "struggle"—a necessity for natural selection and one of the specific principles of creation in the biological
world—which he speculated was the main principle underlying all
existence. This led him to build all of his theory on struggle, and
this focus on one side of the reality of biological diversity served
to increase his influence over others.

The second factor was the inadequacy of Christian scholarship
in explaining advances in the field of geology. The idea that Earth
had slowly changed over millions of years and had taken the form

it has today was not accepted in the beginning. When relatively positive evidence was presented to prove that mountains, rivers, lakes, seas, forests and deserts had gone through many stages, this made it easier to accept the thought that plants and animals had also slowly developed and evolved from simple forms over a very long period of time.

The third factor was that the oppressive behavior of the Christian establishment during critical historical periods which extended back to the Inquisition in the Middle Ages broke up the relations between scientists and the Church. The misinterpretations of Christian theologians, which resulted from literal interpretations of the Biblical description of creation, was insufficient for a proper understanding of scientific developments, and thus inherently contradicted man's intellect and logic according to the needs of the time.

The fourth factor was that the context of the hypothesis of evolution made it suitable grounds for comment by Marxist, materialistic and positivistic philosophical movements, as well as making it serve certain fascist and racist doctrines. Another factor was the expectations which were caused by rising income and prosperity levels due to social and economic agitation in various sections of English society during Queen Victoria's reign.

Evolution does not just consist of the claim, "Humans evolved from apes," as the public commonly understands it. Even though the biggest fuss is made over the idea that humans and apes differentiated from the same ancestor by lineage split, this is actually only a part of the evolutionary hypothesis. For this reason, it is possible for some people to start thinking, "If God wills, He can create both humans and apes from the same ancestor; or if He wishes, He can bring man into existence from a living being similar to an ape." Nevertheless, the basis of evolutionary thought relies on not only the evolution of humans by itself, but also the evolution of the whole universe—the evolution of everything, living and nonliving, by coincidence, without needing a Creator. Simply, then, the debate over humans evolving from apes is the reduced topical extent of the discussion in public. Of course, God can create any living being in any shape according to His will. However, by trying to impose actual mechanisms (isolation, mutation, adaptation and natural selection) on the emergence of human beings, evolutionary thought claims that the laws of nature—blind, mindless and unconscious forces which came together by coincidence—created all living creatures.

According to evolutionary thought, then, the chain of coincidences that started with the Big Bang sequentially followed one after the other: they formed all of the galaxy systems, the star islands, billions of stars, the Milky Way, the solar system, and the Earth—and the most convenient conditions for life on Earth for all living beings. Such thinking asserts that there is no need for a Creator since the formation of all these mechanisms arose by itself without any knowledge, willpower, power, intention and purpose. Thus, having such a structure, evolution completely works as a tool of atheism.

Most evolutionists claim that the idea of evolution is a theory, but for some of them it is almost a definite law. Indeed, evolution is an idea that cannot go beyond being a hypothesis in its form. No other hypothesis has been discussed for such a long time in the history of science. A hypothesis, proposed to explain any event, becomes a theory—or not—after it has been tested by many experiments and observations, depending on whether the results confirm it or not. If the hypothesis becomes a theory, then after it is used for a while, it may either become a law and general principle—due to the power of its explanation—or it may be abandoned due to its inadequacy.

Those who believe in evolution have conducted many experiments to confirm their thoughts, and they have made elaborate comments about countless observations they have made. However, they have not found sufficient explanations or convincing proofs beyond a certain level to support their idea; thus, their theory is left incomplete and insufficient. In reality, we have no idea how conditions were on Earth in the beginning. The first moment of the creation of the universe, and the amazing, miraculous events that happened afterwards are not known either. Ideas are made up based merely on some properties of the present elements and rocks, in the belief that these are accurate indications of history. In addition, evolutionists—who have described the conditions on Earth in the beginning through their own desires—chose the basic characteristics of the artificial Earth in such a way that it could cause amino acids, and thus proteins, to emerge by themselves, and then they planned the Earth's atmosphere according to their dreams. However, it was found out in these experiments performed within the conditions considered to have existed on Earth at that time, that even synthesizing one protein molecule, which is the minimum precondition for life to emerge, is not possible. Besides, there are so many studies which show that the initial conditions of the Earth and atmosphere did not take the forms which evolutionists claim. As the reader will see from the answers to the questions below, despite the fact that it has continuously been disproved by experiments, evolution has persistently been defended as a theory. Never before has a theory which has been exposed to so much refutation been kept in the spotlight at such a level so as to distract so many people. In fact, rather than being a law or even a general principle, evolution can only take its place as a hypothesis in scientific discussions.

Evolution is not a theory, nor has it any relation to science. The definition of science, its characteristics, and the criteria of being "scientific," have been explained in detail in countless books of epistemology and the philosophy of science, and famous philosophers, such as Kuhn, Popper, Lakatos, and Feyerabend, have discussed the structure of science. In short, science deals with events that are determined by repeated experiments or by clearly measured and evaluated data or criteria. Speculation, only, can be made about events which happened once in the past so that their repetition is impossible; thus, scientific criteria are not applicable in the search for the true nature of those phenomena.

According to Karl Popper, in order for a theory to be scientific, it has to give us the opportunity to prove its fallibility through scientific experimentation. For example, physics is a true science because it makes predictions about events, which can then be disproved in principle. In other words, the possibility of demonstrating fallibility is not a weakness in a scientific field; rather, it provides a great advantage in terms of verification, and it builds a strong base for studies in that field. Also, it provides opportunity for separating error from truth, and the relative applicability of a theory concordant with "nature" becomes observable. To Popper, evolution is not scientific like physics and Marxism, since it has the significant deficiency that phenomena are always being observed and interpreted only in order to verify it.

Thus, the attribute of being falsifiable earns the merit of being a fundamental concept in science, and we may call this "the criteria of defining the limits." On one side, then, there are theories which can be disproved through experimentation; on the other side, there are groups of theories which are unclear and which do not allow verification through testing. Those in the first group belong to the field of science, while those in the second group belong to the field of metaphysics. Evolutionary theory belongs to the second category. Popper emphasizes the fact that evolution is not a scientific theory, for Darwinism is not a testable scientific theory, but a metaphysical research program; in terms of testable scientific theories, evolutionary theory is very rough and open to all kinds of criticisms. He does not think that Darwinism can explain the origin of life.[3]

Philippe Janvier says that a metaphysical theory can be true, but a weighty deficiency arises. For him it is practically impossible to test evolutionary theory.[4] The reason is that if the history of life on Earth, its emergence and development, is thought of as a frame by frame, true-life movie, it is not possible to rewind the movie and watch it again from the beginning. Since it is argued that evolution occurred over a long (geological) time scale, it is not feasible to test it through experiments or observations. For this reason, it is not possible for the natural sciences to disprove it. A theory which does not provide the opportunity to disprove it, and thus is not falsifiable, does not have the qualities required to be accepted as being scientific.

At least we can say that we are faced with a shameful and bewildering situation. Just thinking or claiming that evolutionary theory is scientific, on the other hand, does not simply make it suitable for scientific testing. It cannot be observed, derived or measured. However, its advocates want it to be considered a certain and provable fact about both the beginning and progress of life. In such a case, every self-respecting scientist who has any confidence at all would, and should, ask for concrete evidence about it. Yet, Russian biochemist Alexander Oparin aptly says, "If we are after proof, we will never be able to attain it."[5] In his opinion, it is not possible to find evidence in chemistry or physics concerning the biological formation of the first creature.

Yet, if we cannot prove evolution by scientific methods, then likewise, we cannot prove the opposite. This is undoubtedly a reasonable assertion. The same thing is also true for all other theories which contradict the firm conditions which are set for the scientific method. For, as discussed above, in order for a theory to be accepted scientifically, it has to be open to falsification. In other words, a theory has to be tested for its truth or falsehood to be proven. A set of ideas that cannot, in principle, be falsified is simply not scientific. For example, Newtonian physics is a theory which *can* be falsified since Newton's Laws are open to experiments on their validity and can be tested. Conversely, however, it is impossible to determine whether evolutionary thoughts are scientific truths or not. Even Darwin himself understood this essential truth. In a letter he wrote in 1863, he admits that if we descend to details, we can prove that no species has changed (i.e. we cannot prove that a single species has ever evolved), nor can we prove that the supposed changes are beneficial, which is the groundwork of the theory.[6]

Therefore, since evolution cannot rely on scientific observations, it has to be a matter of personal belief. The best thing to be said about evolutionary theory is that it represents a belief which is neither provable nor falsifiable and which is shared by many people about how life evolved. Certainly, everyone is entitled to their own beliefs, theories and personal opinions. However, evolution advocates claim that evolutionary theory is something beyond being a matter of belief. According to them, evolution is a clear fact, even though it cannot be proved, and they do not show any tolerance towards counter-arguments about the fundamental doctrines of evolution.

Perhaps some will consider that this is not a tragic situation, but it is still important to reflect for a moment on the evolutionist's brutal attitudes while asserting their ideas, and their extreme intolerance toward alternative opinions. Such attitudes remind everyone of a very common pattern of behavior which has been witnessed, unfortunately, since the beginning of humanity. Today, evolutionists are "devoted believers" with all their faculties: they were baptized in natural selection, they started spreading the good news, and they began distributing the message to others, so that they, too, would accept Darwin's doctrines.

2

Arguments on the Origin of Life

ARGUMENTS ON THE ORIGIN OF LIFE

T he crucial questions are how living beings arose on earth, how many of them came to life, and how the over two million animal and plant species which have been identified spread throughout the world (It is estimated that as many as 10–30 million species may exist). Significant improvements have been made to the explanation and understanding of biological life in the fields of anatomy, physiology, genetics, biochemistry and cytology, and in the area of health and nutrition. Thus, biology is predicted to be the pioneer among the branches of science in the twenty-first century. For example, it is expected that the most difficult problems, such as cancer, AIDS and genetic diseases, will be solved using biotechnological methods.

In spite of all these advances, the creation of the universe, the Earth, life and humans all appear likely to remain secrets which exist beyond the study of science, and which exceed the limits of science. The claims about the first emergence of creatures will not go beyond being speculative arguments as no one ever witnessed these events. It is not possible to repeat the first creation through scientific experiments and observation as one might replay a video recording over and over in slow motion. The reason behind this is twofold: the impossibility of recreating or regenerating the very first living beings mentioned above; and the impossibility of designing or establishing a model of the actual physical and chemical conditions which were present during the process of the first creation.

In addition to this, the human intellect—which is curious about every event, questions everything, and aims to find an explanation for all phenomena—continues to discuss ideas about the origin or emergence of life under four categories. Three of those are the outcome of the human mind and intellect, and the other is based on divine sources. Simply put, there is no logical way other than these four in which to consider this subject matter.

1– The most nonsensical and completely discarded argument regarding the emergence of life is the idea of "**abiogenesis**"—the claim that life emerged from lifeless material having no biological origin whatsoever. According to this view, which is abandoned today, living beings arose from non-living matter all by themselves. The ancient Greek philosophers who were the very first advocates of this theory—namely Aristotle (384–322 BCE), Thales (sixth century BCE), Anaximander (610–545 BCE), and Xenophanes (560?–478? BCE)—believed that living beings emerged from non-living matter through so-called *generatio spontanea* (spontaneous generation), entirely on their own. According to this line of thought, plant bugs arose from dewdrops; frogs rose from marsh mud; and flies emerged from decayed wood and organic material. These ideas found a number of fans in Europe well into the Middle Ages and even up to modern times. In the seventeenth century, many biologists—such as a well-known Belgian, Dr. Jean Baptiste van Helmont (1580–1644), a British scientist, Needham (1713–1781), and a French researcher, Pouchet (1800–1872)—supported the theory of abiogenesis and conducted experiments related to it. In fact, Van Helmont's thesis that a mouse would arise from a dirty shirt and wheat in twenty-one days remains an interesting historical claim relating to this idea.

The argument—which started with the idea that single-celled living beings were produced in rich organic solutions, such as boiled straw and meat broth—was vanquished by the experiments of Francisco Redi (1626–1697), Louis Jablot (1645–1723), Spallanzani (1729–1799) and ultimately, by Louis Pasteur (1822–1895). Upon completing his experiment, Pasteur concluded that "obtained results show that even microscopic living beings cannot be formed without ancestors resembling them." After it was proven that single-celled living beings which reproduce in straw and meat broth are actually emerging from spores which are transmitted from air to water, and that the maggots which reproduce on flesh actually emerge from the larvae of the flies which leave their eggs on meat, no one continued to claim that a living being could emerge from non-living matter.

On the other hand, naturalistic theory, which results from the hypothesis of evolution and which we will mention below, is actually a kind of "modern abiogenesis." Even after it was understood clearly that it is impossible for a living being to emerge from non-living (inorganic) matter, either on its own or coincidentally, efforts to search for a way to form a living being out of non-living matter in various fashions through gradual accumulations over time have been designated as evolution."

2– While abiogenesis is an idea which originated in ancient times, "**cosmic theory,**" the second idea, became known especially after advances in astronomy. According to this view, dust pieces in space, and the organisms on meteorites, such as bacteria, were the first sources of life on Earth. It is thus claimed that even in the cold, oxygen-free and lethally radioactive environment of outer space, some of the organisms on meteorites and asteroids reached high temperatures due to the friction which resulted when they entered the atmosphere at high velocities, and they finally reached the Earth to become the source of life. This theory, almost abandoned today, has been criticized from various perspectives, as no reasonable proof to support it has yet been found. According to current scientific knowledge, it seems impossible for any microorganism to travel for so long in space through such strong radioactivity, to survive in spite of the extremely high temperatures caused by friction upon entering the atmosphere, and to reach Earth safely under such difficult conditions.

Besides, even if we allow that a living being somehow reached Earth from outer space or from another planet, another crucial question—how did such a living being arise on that other planet?—still remains unanswered. Related to this is an interesting example where certain scientists first claimed that a meteorite, which appeared to have a microscopic formation of worm patterns on it, had broken off Mars and fallen to Earth. In the beginning, it was even suggested that the worm patterns were structures which had formed as a result of bacteria activity, or a type of pathogenic fossil. However, more recent studies showed that the patterns were actually completely inorganic structures which were formed at very high temperatures that would not allow such kinds of life.[7]

3– The third argument is "**naturalistic theory**." Even though it resembles abiogenesis at first glance, it is actually different from it. This is because while an original living being arises directly from non-living matter according to the idea of abiogenesis, according to naturalistic theory, a simple living being emerges first, and then this simple living being forms a developed organism by evolving over a very long time. The idea of evolution is considered naturalistic, for in order to build a foundation for their arguments based on the first two ideas explained above, evolutionists use biological processes most often to explain their materialistic opinions. For this reason, this theory seems as if it is scientific at first glance—and since it apparently gathers evidence from nature, it is considered "naturalistic." Within naturalistic theory, there are two different underlying hypotheses:

    a. *The Hypothesis of Autotrophy*: According to the hypothesis of autotrophy, the first living being arose coincidentally by itself and thus, that living being had to make its own food by itself since there was no food in the initial environment on Earth, for there was no other life yet. Because of this, such a living being had to have the ability to synthesize its own food from inorganic matter using sunlight (photosynthesis) or some sort of chemical matter (chemosynthesis). In other words, according to the hypothesis of autotrophy, it was necessary for the first living being, which effectively had to make its own sustenance, to have very well-developed enzymes and mature synthesis mechanisms. However, the complexity of the biochemical reactions related to the synthesis of organic matter has always been a problem for this hypothesis. This is because it is so unlikely that a system that requires a perfect plan and program could be established suddenly, by itself, and be instantly ready for the production of complex molecules, such as sugar, from solar energy—or immediately able to convert them into bigger molecules, such as starch and cellulose. The fact is that it is more reasonable to accept the emergence of such a complicated living being, with such excellent synthesis mechanisms, as being possible only through the will of a Creator Who has infinite knowledge and power. In other words, no strictly scientific credit can be given to coincidence—and thus, materialists have had to abandon the hypothesis of autotrophy—one of the supposed pillars or presuppositions of evolution.

b. *The Hypothesis of Heterotrophy*: According to the hypothesis of heterotrophy, the other conceptual foundation or prerequisite of evolution, in order for a first primitive living being to arise, inorganic matter in lifeless nature evolved for long enough in favorable conditions—presuming first that lifeless inorganic molecules (i.e., amino acids and proteins) emerged, and then the first primitive cells, complex cells, and primitive plants and animals—so that complex plants and animals could eventually come to life randomly by the coincidental compounding actions of those inorganic molecules.

This assumption of evolution, which seems as if it explains everything at first glance, was quickly raised to the level of a theory. Since it supported Marxist and materialistic views, it was presented as if it were a proven law that had been confirmed by repeated experiments. Thus, rather than a biological postulate, the idea of evolution became an ideological doctrine—even a religion for some. As will be explained later on, evolutionary theory posits coincidental chemical reactions and random occurrences of mutations and natural selection. This idea surely rejects the perfection, the planned creation, in the universe—and thus, the Creator.

4- The last one is not an argument but the belief in the reality of **creation**. It is a belief that all living and nonliving beings were created with an excellent plan and design by a Creator Who has infinite knowledge and power. As an obligation of this belief—which is the foundation of the divine religions—it is understood that nothing is purposeless. In addition, it is believed that the first cause of everything is the Creator, One God, Who sees and takes care of every living being at every moment. Thus, He created and designed all creatures with the most suitable organs and senses, and He prepared them for conditions on Earth in a most ideal way.

The reality of creation is not an opinion; rather, it is the knowledge which is put forward by all divine religions, and which is agreed upon and confirmed by all prophets and holy books which have delivered the revelations to humankind. The information about creation has been conveyed to people in a special form, called "divine inspiration," by the prophets of God. Divine inspiration is a way of stating divine declarations, so it cannot be considered a subject area of science, which is limited by experimentation related to our material world and sense organs. It is the truth that can be felt by intellectual, heartfelt and spiritual experiences and observations of a person. For this reason, it cannot be constrained within the confines of science; yet, to make divine inspiration clearer for human minds, science could offer some evidence that opens horizons, answers doubts, and brings the concept of creation closer to rationality. In addition, it could help to manifest the impossibility of disbelief by providing evidence about the fact that creatures would not, in fact, have come into existence without a Creator.

The most powerful refutation of the theory of evolution and materialistic philosophy came from Bediüzzaman Said Nursi in

Turkey. Nursi concisely pointed to this issue first in Al-Mathnawi al-Nuri which he wrote in Arabic in 1923–24. Later in the 1930s, he expounded on the topic in a treatise he called "On Nature" which was published in his *Lem'alar* (The Gleams).[8] With impressive argumentation Nursi gives convincing examples in this treatise against the theory of evolution without mentioning it, for Nursi knew atheism was the underlying philosophy. The principles that Nursi proposes in his article can be thought of as an enhanced prescription for modern versions of abiogenesis, and his perspective is useful and important in helping us to understand this issue. First, we should review, in brief, the only four ways in which the existence of living beings and their perfection and order is considered possible:

   a. Living beings come into existence by the random influences of forces, called "causes," such as air, heat, light, damp and the forces of attraction in atoms.
   b. Living beings come into existence from nonliving beings, by themselves.
   c. Everything is created by "nature."
   d. God creates everything.

The impossibility of the first path, wherein "causes make living beings come into existence," is ultimately explained by probability calculations. It has been calculated that the probability of combining 40,000 atoms in a protein molecule in a particular design is 1 out of 10^{160}, and the time that would be required for this process to be completed is 10^{243} years. Thus, we face a separate problem regarding whether the astounding odds against such a coincidence make it even worth discussing. In addition to that, the miraculous flow of all complex events without any malfunction or confusion in a living body can never be attributed to unconscious or mindless causes. A creature having measure and order must surely arise through planned organization and a great deal of knowledge; otherwise, it would be impossible for innumerable causes to come together with agreement in a living being in the most suitable amount, time, place and conditions in order to form such a living being. The uncertainty of the emergence of a living being by the random influences of forces will also be explained in detail later using "probability calculus."

The second path implies that in order for any living being come into existence by itself, all the pieces forming the whole being must know every detail of this whole being, and they must also be able to intercommunicate in order to come to an agreement about which function of the whole body they each will assume, following which each one then has to take on its respective role. Thus, every single atom that works in an organism has to have precise knowledge about it so that the entire body can work and continue its functions properly. As is well known, since matter has a tendency towards disorder rather than order, it can neither form an orderly structure nor can the necessary energy for the system to function be conserved. According to the second law of thermodynamics, while it is only possible for a living system to continue its existence by the conservation and management of its matter and energy in a controlled fashion, disorder and dispersal occur by themselves—they do not need any external intervention. In other words, the Creator taking His will and willpower out of that living being is sufficient for the system's matter and energy to lose order. In fact, even the systems that we think of as having their order broken by themselves actually undergo disorder according to the will of the Creator. Therefore, there has to be a source of infinite knowledge and power, for this is a fundamental requirement for preserving the order of matter and energy in living systems and for resisting disorder.

The third path essentially refers to nature as the universe itself, wherein nature consists of components such as birds, trees, stones, insects, bacteria, flowers, flies, and so on. If we claim that each component is created by nature itself, then if we take these components away, nothing at all should be left of "nature." Seeing this impossibility of attributing existent things, animate or inanimate, to nature and seeing all the rules or so-called "laws of nature" which are necessary for the subsistence of nature itself, all existent things must inevitably be attributed to a form of agency which creates the laws which are beyond and outside any one of the particular components of nature. Thus, the existence of interactions governed by natural laws presupposes a Lawgiver. In other words, nature is a work of art, but not the Artist—and the face of the Earth is a painting, not the Painter Himself—for there must be an Artist Who creates the painting that we call "nature."

Seeing the impossibility of these three paths, the fourth path, God's creation of everything with His infinite knowledge and willpower becomes the only reasonable way to explain the origin of life.

Some metaphysical questions always arise after new discoveries in various fields of science. This shows that the ostensible categorical differences between philosophy and science are actually artificial. As a matter of fact, Herbert Spencer (1820–1903) comments on this point in his text entitled "First Principles," from his collection called *Synthetic Philosophy*:

> [...] At the uttermost reach of discovery there arises, and must ever arise, the question – What lies beyond? As it is impossible to think of a limit to space so as to exclude the idea of space lying outside that limit, so we cannot conceive of any explanation profound enough to exclude the question – What is the explanation of that explanation? Regarding Science as a gradually increasing sphere, we may say that every addition to its surface does but bring it into wider contact with surrounding nescience. There must ever remain therefore two antithetical modes of mental action. Throughout all future time, as now, the human mind may occupy itself, not only with ascertained phenomena and their relations, but also with that unascertained something which phenomena and their relations imply. Hence if knowledge cannot monopolize consciousness – if it must always continue possible for the mind to dwell upon that which transcends knowledge; then there can never cease to be a place for something of the nature of Religion; since Religion under all its forms is distinguished from everything else in this, that its subject matter passes the sphere of [the intellect] [experience].[9]

Philosophers who speculated on the existence and the beginning of life could never avoid debate with others, for such speculations have a tendency to become some sort of a worldview. There was a very serious debate between two French zoologists, Cuvier (1769–1832) and Lamarck (1744–1829). At first glance, it seems that the dispute was about the difference between the fossils of vertebrates and invertebrates, but it was actually not limited to this; rather, the debate had an ideological side. Cuvier, advancing his idea of catastrophism (recreation after certain extinctions), predicated his thoughts on the Bible. To Cuvier, who saw the discontinuity among various species, it was impossible to claim the occurrence of transformation from one kind of species to another. Contrary to this, Lamarck thought that species could change "as a result of time and conditions"—namely, that the transformation from one kind of animal to another could occur.

In what was essentially a first debate about the idea of transmutation, which would form Darwinian evolutionary theory in the future, Lamarck tried to explain "the transformation of species" based on the idea that some species emerged from others by virtue of a hypothesized "inheritance of acquired characteristics," an idea which would later be renamed simply as "Lamarckism." As evidence, he claimed that giraffes could arise from a mammalian animal as big as a goat, for instance, as the result of efforts to reach from the lower to the upper branches of trees, essentially stretching their necks over the course of thousands of years. Many believed in the idea of the inheritance of acquired characteristics at that time, and people would continue to believe in it for a very long time. Indeed, Darwin himself would later embrace this notion. However, this hypothesis would eventually be abandoned in the twentieth century due to improvements in genetics and cytology. It became simple information which almost everyone knows today that acquired characteristics cannot be transferred to future generations unless they are transmitted to the genes. Just like the failure of Weissman's unfortunate attempts to obtain a tail-less mouse by cutting off the tails of mice for a couple of generations, the classic examples of the falsehood of Lamarckism include the continuity of uncircumcised children born to Muslim and Jewish children, even though they have been circumcised for hundreds of years; and the unchanging size of Chinese females' feet, even though their feet were purposefully narrowed in childhood for generations in times past. We now consider the changes that occur strictly in the phenotypes (in the physical appearance or manifestation of a living thing) but not in the genotypes (in the genes of a living thing) to be modifications only—and it is well known that modifications do not have any importance for evolutionary theory today.

When Lamarck died, Cuvier wrote *Elegy of Lamarck*, which was a kind of academic criticism rather than representing admiration or commendation. The attitudes and feelings of Cuvier toward Lamarck can easily be ascertained by reading a couple of passages from this elegy:

> "[Lamarck's evolution] rested on two arbitrary suppositions; the one, that it is the seminal vapor which organizes the embryo; the other, that efforts and desires may engender organs. A system established on such foundations may amuse the imagination of a poet; a metaphysician may derive from it an entirely new series of systems; but it cannot for a moment bear the examination of any one who has dissected a hand, a viscus, or even a feather."

Indeed, Cuvier basically blamed Lamarck for not examining any organism anatomically. Even though they were buffeted by Cuvier a little bit, Lamarck's ideas—and especially his theory of "transformism"—had a certain philosophical viewpoint. In any event, Lamarck was separated from his predecessors in this regard. For example, before Lamarck, Maupertius (1698–1759), a cosmologist and mathematician, affirmed the idea of biological change (mutation); on the other hand, he also tried to prove the existence of God by aiming to reach a unique and simple principle that combines all of the laws in the universe. Lamarck also attempted to introduce the hypothesis of transformism, or trans-mutation, by offering evidence, or what he called, "the pieces of evidence." Further, he aimed to propose a systematic correlation between the fossil records and living invertebrates, and the taxonomy of the 150,000 invertebrates and 15,000 vertebrate species which were known at that time.

So, what happened to the idea of transmutation after Lamarck died? Needless to say, Cuvier did not support Lamarck's theory. Since he was predominant not only politically but also scientifically (he was a university rector and had close links with the political circles), he was the one who organized research groups and appointed students with respect to this purpose. Then, fifty years later, Darwin's advocates took control of everything, and they could not bear the idea of someone else (i.e., Lamarck) having postulated or circulated of the notion of transmutation other than Darwin.

But Lamarckism existed and developed in spite of Darwin, and it even came back as "Neo-Lamarckism" in the United States at the beginning of the twentieth century. At that time, most of those who believed in Darwin and in natural selection as a totally new concept revived Lamarck's idea of transmutation, and they accepted Darwinian transmutation as an improved version of Lamarck's transmutation. This comment was first expressed by one of Darwin's professors, the British geologist, Sir Charles Lyell. According to Lyell, the only thing that Darwin did was to improve Lamarck's theory since evolution, which was described as a transition from one species to the other, is the fundamental notion behind transmutation.

In the twentieth century, especially Albert Gaudry (1827–1908), a professor at the Natural History Museum in France, and later, his student, Marcellin Boule (1861–1942), were pioneers of paleontology. Some other paleontologists in different countries, such as Richard Owen (1804–1892) in England, and Cope and Marsh in the United States, were also very ambitious to improve on this theory, and Cope would become the one who would cause the ideas of Neo-Lamarckism to be circulated widely.

3

Evolution and Creation:
What Do They Promise?

EVOLUTION AND CREATION:
WHAT DO THEY PROMISE?

When we compare the belief in evolution and the belief in creation as to their advantages and disadvantages, we clearly see how destructive the belief in evolution is for human nature and social life.

1a. The causes of change on which evolution is based are coincidences, random chemical reactions, and casual mutations.

1b. However, according to the belief in creation, none of the events in nature happens by chance, be it randomly or coincidentally—not in the least.

2a. According to the hypothesis of evolution, biological events and processes emerge only by means of material cause -effect relationships.

2b. Conversely, according to the belief in creation, causes cannot be denied, but they are just the reflections of God's operations—thus, material causes should be sought as part of our efforts to understand the ultimate reason behind natural phenomena.

3a. According to the hypothesis of evolution, natural selection is a hard struggle in which the strong survive and the weak die. Spiritual realities—such as affection, compassion and dependency—cannot even be mentioned, let alone be accounted for. Instead of cooperation and altruism, the dominant, inherent mentality is that of thinking only about ourselves; thus, the idea that "I don't care if others die of hunger so long as my own stomach is full," is valid and reasonable according to the belief in evolution.

3b. However, according to the belief in creation, natural selection is not merely a struggle to survive that is lacking in affection and compassion. Even though struggle in the competition to survive is evident, cooperation, solidarity and compassion—veiled by divine compassion—remain indispensable in this struggle. Every event in nature has a wisdom and purpose that we do not know. For example, it is because of the need to keep the balance in the ecosystem that weak and unhealthy animals become food for strong ones. In this way, the surface of Earth does not become a dumping ground, there is enough space for new generations as the old and sick are removed from the environment, and the continuation of the food cycle is assured.

4a. According to the hypothesis of evolution, the laws of nature, which make life arise from the elements of inorganic matter, do not have an intellect, consciousness, knowledge or power; and thus, it is pointless to search for an Authority, or Artist, since these "laws" apparently do not have a larger purpose. That is because the artist of the laws of nature, according to such thinking, is nature itself. This logic supposes that any living mechanism is built by the coincidental activities of the atoms and molecules which form it, and such a living system operates by itself; therefore, there is no need to look for an Artist behind any natural mechanism or its function.

4b. In contrast, according to the belief in creation, the laws of nature are not the Artist but rather, the Artist's products—works of art that do not have an intellect or consciousness. There is indeed a Creator, Who establishes the "laws of nature," protects this system by operating it according to His orders, takes preventive measures to protect it, and governs those laws under which it functions optimally. Since everyone admits that a desk or an automobile cannot come into existence coincidentally by itself, and that a craftsman who makes such things will need to be found, likewise, it is not possible for a cell which is millions of times more complex than a desk, or for a human brain which is trillions of times more complex than an automobile, to come into existence coincidentally without a Craftsman.

5a. Some materialists may attribute a hidden consciousness and intellect to atoms and molecules when they observe their unfailing functions and the faultless program of each atom in regard to its determination of where and when to move within those biological processes—and they might even consider atoms and molecules to be sentient creatures with willpower.

5b. On the other hand, the belief in creation does not attribute any knowledge or willpower to atoms and molecules. Atoms and molecules are particles without any will or consciousness, which strictly obey the commands of their Creator and fulfill their duties without any resistance or imperfection in their compliance.

6a. Even though it seems to be an assumption about biology, evolution has actually been the basic philosophical substrate for materialism and atheism for one and a half centuries. It has unhesitatingly been used to oppose belief in God under the appearance of being scientific. For this reason, evolution should be considered not as a scientific theory but rather, as a belief which is contrary to religion.

6b. In contrast, the belief in creation is a complete worldview based on religious resources. There is no difference between evolution and creation with respect to the measures for being considered "scientific" today. The only difference is that evolution is an atheistic worldview and creation is a theistic worldview.

7a. The particular language of those who support the ideology of evolution will easily be recognized in their explanations of natural phenomena: for example, "a living being has developed," "developed by evolution," "its legs disappeared in time," "gained by adaptation," "emerged by natural selection." As these phrases clearly reveal, the underlying claim is that there is no need for a Creator since the laws of nature themselves somehow "create."

7b. However, those who believe in creation have also developed their own distinctive phrasing; for instance, "was created in this form," "was created in the best form," "was planned and designed in the most perfect way." As this type of language reveals, the Creator of the living being is implied through an emphasis on harmony and planning, and through the focus on the inherent program, organization and systems of natural forms and processes.

8a. For evolutionists, the presence of appropriate organs in the body and their excellent functions in an organism result from the processes of adaptation and natural selection. Thus, it is not reasonable to search for purpose or wisdom behind these structures, nor to think about a Creator.

8b. On the other hand, for "creationists," it is accepted that each organ has been purposefully made by the Creator for a specific aim, according to complex divine wisdom. As a matter of fact, since organic factories like the cell, and complex organs like the eye, are perfect systems, it is not possible for them to transform from a defective or partly-developed structure to a fully functioning form strictly by evolution—for this itself would presuppose a certain purpose. Does it really make sense to one who has reason and perception to believe that two particular parts of the body could develop coincidentally in order to form an eye or ear, consciously and decidedly, while nothing was present in their place in the beginning?

9a. According to the hypothesis of evolution, there is no point to putting humankind in another or higher position by separating it from other living beings. After all, humankind is just a little different from the ape species; in other words, it is just a slightly more intelligent animal. From this perspective, humans can simply follow the most basic animal laws as other animals do; thus, they tacitly obtain permission to abandon their ethical and humane values.

9b. Yet, according to the belief in creation, humans were deliberately and purposefully created to be distinctive from other creatures on purpose, so that they might know and acknowledge their Creator. That is why they have been given such faculties as the mind, conscience, heart, and soul, as well as other attributes which enhance their perceptive and cognitive abilities. Being the highest of all creatures, humans should demonstrate that they are different in nature from animals by recognizing their Creator and by following the ethical principles which their Creator orders in recognition and gratitude for being created in such an honorable position, and to demonstrate their understanding of the purpose of his existence.

10a. The most widespread result of having faith in evolution as if it were a religion is that as a worldview it is influential and causes new discussions in almost all fields of science, from astronomy to sociology, and from physics to psychology. Views such as Marxism in economics and Freudianism in psychology, along with evolutionary theory, have become allies that attack the same target. People with an understanding that nature is Owner-less, and who do not admit that they will be called to give account of everything they have done, are likely to cause the exploitation of the environment as well.

10b. On the other hand, a faith which is based on one of the divine religions, and the consequent worldview, will also be reflected in all scientific endeavors. One's views and evaluations of nature in different fields of science will then highlight the principles of one's ethical values and the substance of one's conscience, and the scientific research which emerges from this perspective will be beneficial to humanity. Protecting the environment and all life, and looking after both humankind and nature as trusts from God, will be the results of this point of view.

In general, serious discussion about the ideas of evolution and creation is rarely conducted among scientists whose thoughts have been shaped by these two views. This is because the subject matter exceeds the limits of science in that it has a special feature that requires interpretation. If the subject matter were within the limits of science—that is, if experiments and observations were performed about evolution and creation—then there would not be such a problem. For example, there is no difficulty with physics problems that are included within the bounds of science, such as the law of gravity, the calculation of the expansion of metals, or the lifting force of water or air pressure. However, it is easily witnessed that debates are raised and the subject matter is scrutinized as a worldview or belief even in physics when it comes to any topic below the atomistic scale—for instance, quantum mechanics, antimatter, existence versus nonexistence, and so on.

The reason behind this is that humans feel the need to believe in a value system and connect to it as a necessity of their nature stemming from their creation. In order to satisfy the feeling of belief and connection that is present in their hearts and consciousness (and it is to be hoped that they come to know the books of the Universe and the Qur'an or they are left wanting in this regard), humans either view nature as the "art of the Creator," or as the natural result of evolution. At this point, the situation of a scientist has particular significance.

If we think about the fact that all scientists are raised in their own family environment and in their own society, with certain values and principles, we cannot expect any one of them to be absolutely objective. In other words, how reasonable or realistic would it be to expect a scientist to leave his or her belief entirely out of the laboratory?

A scientist who performs his or her studies from the perspective of faith—a "scientist of faith"—will always point to the Creator while interpreting his or her studies. Another one who sees everything from an atheistic point of view will interpret results according to materialist and positivist philosophy. Yet, though both of them are clearly considered to be a belief and worldview, arguments and ideas that are supposed to be discussed easily in a democratic country are presented in an offensive and aggressive manner without showing any sign of respect, tolerance or patience towards those of opposing ideas.

First of all, it is contrary to the methods and discourse of modern science to present evolution as if all of the issues about it have been resolved, or as if it has been proven. As a matter of fact, the evolutionary theory that was first proposed by Darwin always raised reaction from very broad segments of society. Nonetheless, evolutionary theory managed to quiet initial reactions—even in the face of the doubts instilled by religious ideologies within the Christian Church and particular interpretations of the Bible—partly because of the inadequacy of Christian resources in giving satisfactory answers to those early suspicions. Thus, it slowly took its place in the scientific community and began to give the impression that it had gained a scientific identity. In the meantime, scientists of faith stayed silent because they feared the accusation of being unprogressive and outdated in the oppressive atmosphere that was created by evolutionists, who actually used the public's interest and trust in science to advance their own vested interests.

The arrangement of ape figures that is supposed to picture the gradual transition from the creatures similar to apes to humans took its place in textbooks as if it were evidence from nature. These figures purported to demonstrate how ape-like creatures straightened up on their two feet from four feet, developed bulging lower jawbones and prominent foreheads, and molted their hair. In addition to this, a debate was ignited with regard to which particular animals might be human ancestors, amid the branches of pseudo-lineage trees that presumed to show the coincidental "derivation" of all animals from one another, from single-celled organisms to mammals.

Yet, even though evolution was presented as a proven law in many countries until the 1950s, the discussions between the advocates of evolutionary theory and the supporters of creation really heated up after it was understood that some of the fossils which had been presented by evolutionists were fake and deficient.

In such an atmosphere—and while it was very hard to express any direct thoughts against evolution—Watson and Crick discovered the DNA molecule and presented its structure in 1953. With the discovery of the structure of DNA's double helix, the perfect structure—of a program and process that does not allow coincidence in the cell and thus in any living body—became widely known and so the belief in creation was strengthened once again. At the same time, some scientists developed new approaches that disprove evolution to show that creation, the reality shown by religion, is in complete agreement with the realities which are discovered by the methods of modern science.

In turn, each new research venture, and every discovery in the fields of molecular biology, genetics, biochemistry, cytology, embryology and physiology, actually shows how little information scientists have about the phenomenon of life itself—as each new discovery displays the marvels of complex life even more deeply. As the realities which are proven and demonstrated by countless complex and perfect biological structures—which cannot possibly have arisen through coincidence—accumulate, they become an impassable obstacle for the notion of evolution. Believers in creation thus managed to finally escape from the oppressive atmosphere which was created in the early years of evolutionary ideology, and year by year, the number of studies that show the deficient and false aspects of evolutionary theory only increase. Pro-creation institutions, such as the Institute for Creation Research in the United States, have been established. Thus, the objections to evolution by scientists who believe in creation have been strengthened over the past thirty years, while the evidence supporting evolution has been increasingly, and correspondingly, weakened.

This situation has caused great discussion in many institutions in the West, and the idea of teaching both schools of thought as the Philosophy of Biology has started to make headway. Similarly, in countries like Turkey, various viewpoints have been added to high school curricula and textbooks since 1980, but this objective approach has consistently disturbed some advocates of evolutionary theory. Because of this, they have tried to estrange the idea from its true purpose by raising objections to teaching creation, such as the idea that religion may interfere in the public sphere; they have politicized the issue in a progressivism vs. reactionaryism and anti-modernism conflict, where belief in creation was purported to stand for the latter. We look forward to a near future where all kinds of thought can be freely supported and no one is persecuted for their ideas. I believe tolerant, bias-free discussion, in which both religion and science are duly respected and not set up as rivals, will yield synthesized thoughts and brand-new combinations of ideas.

4

Biological Mechanisms in Nature

BIOLOGICAL MECHANISMS IN NATURE

Before we go on to the specific examples from various areas of science and their interpretation as accepted evidence by those who claim that evolution occurred, it has to be said that if evolution occurred, then there have to be some basic biological principles underlying the mechanisms by which such an evolutionary process emerged. In order for the hypothesis of evolution to wear a scientific costume, in other words, such biological principles have great significance and people are mostly misled about this issue.

The most important reason for the success of the evolutionists' presentation of their hypothesis as a theory or law is that they start from the biological principles placed by God in nature; however, they completely misinterpret or misrepresent these principles in a way which is contrary to their purpose. Since their initial points of argumentation are biological principles that everyone accepts to a certain degree, the resultant fallacy, and the defective understanding which results, is perceived as if it were true. Those who first attempted to oppose the theory of evolution were not able to understand this logical trick and embarked on a path wherein they denied some biological facts while aiming to oppose the theory of evolution. But as improvements in research and analytical methods showed the truth of the biological principles, an erroneous conclusion formed in the minds of many that the theory of evolution, which hinged on these principles—albeit with critical distortions—was true.

However, there is no point on denying those basic biological principles, which we will explain below. It will rather be understood that all scientific fields—such as molecular biology, genetics, embryology, physiology and anatomy—point to creation through marvelous order and harmony when those principles are correctly interpreted. The fundamental reason for insisting on the idea of evolution as a certain principle simply because it wears the apparel of being scientific is the faulty interpretation of biological principles—which are simply the declaration of the perfect harmony between the genetic program of a living system and the environment where it lives—as evolutionary scenarios. Simply put, due to their proficiency in covering the tautological suggestions that evolutionists offered from within the triangle of mutation, adaptation, and natural selection, due to their perfect familiarity with the enduring paradoxes, and due to their inclination towards interpreting every result to their own benefit, they have managed to give an appearance of scientific fact to their evolutionary ideas. Below, we discuss the real values of these three biological mechanisms and how evolutionists interpret them

NATURAL SELECTION OR THE FOOD CHAIN OF THE ECOSYSTEM?

Another hard-line approach of those who support evolution is that they consider nature a place for struggle. Nonetheless, our interest in and awe at all points of nature shows how beautifully it was created and how amazingly it is maintained. Millions of varieties of species and their innumerable living members sustain their lives in different latitudes and regions. Each appears to be a small or large component of a system that functions in perfect synchrony. Discovering those biological mechanisms at the macro- and micro-levels that constitute such processes and analyzing them in depth became possible in the twentieth century as a result of improvements in science and technology.

Evolution theorists often interpret the life of living beings in nature in terms of the precondition of evolution by natural selection. Natural selection has a merit up to a certain point, but it is not a fundamental law that is always acceptable. Besides, natural selection indicates a law of creation put in place to provide sustenance for all living beings by means of the food chain, which is the basis of the ecosystem.

When there are changes in environmental conditions, such as extreme temperatures, drought, brackishness of water, infectious diseases, starvation, varied pH concentrations—or in the event that the individuals of a certain species migrate to a different environment—some variations that are neutral or harmless might, in fact, become important and the individuals in possession of them can thereby find more favorable living conditions. Thus, some individuals become superior through the advantage of variations in the new conditions, and their chances of staying alive compared to others might increase. Seen from this perspective, physical and biological conditions carry a duty of being a type of "sieve" for natural selection in such a way that living beings which are qualified to live pass this sieve, while those that are not qualified die by "jamming" in the sieve, as it were.

On the other hand, there is not a brutal struggle in nature where only the strongest survive. Daily, we witness the reflections of mercy and compassion in the operation of cooperation and solidarity alongside competition. Those who see the struggle of a small animal population within a restricted area as the basis for all of life's interactions fall into the error of their inadequate observations. When we pay attention to the overall harmony and order of commonplace mechanisms within a wide range of ecosystems, we see the reflections of enormous mercy in the critical balancing acts which comprise the partnerships, cooperation and solidarity witnessed among and between various beings and species.

Every biologist uses natural selection according to his or her own conception by slightly changing its meaning and rendering it compatible with what he or she believes. Thus, because all people look at things from their own particular viewpoint, natural selection easily becomes a subject of dispute.

As to the definition issued by Darwin in 1859, natural selection is a mechanism for maintaining beneficial variations and filtering out harmful variations. The very first question asked as part of the rejection of this definition so clearly deprived of evidence was this: "Doesn't the notion of selection necessitate the existence of selective willpower?"

The views which first sprouted in the mind of Darwin while he was reading the political economist and demographer Thomas Malthus' *Principle of Population* transformed in the following years into such ideas as:

a. There is potentially a geometrical increase in populations.
b. A state of constant and stable balance is observed in populations.
c. Resources are not infinite; they are limited.
    The cumulative interpretation of these three separate observations was that "individuals within a population have to struggle in order to stay alive." Then,
d. Each individual has a distinctive structure.
e. Most individual variations are inheritable.

The cumulative interpretation of these two last ideas was that "the capability for surviving would differentiate each individual in a population, and this would cause evolution over many generations." While the first part of this sentence was an observation of a normal process that detects the strong varieties observed in nature depending on the potential richness within species, the second part of the sentence, however, is nothing more than a judgmental phrase predicated simply on a well-intentioned procrastination but impossible to prove through experimentation.

Natural selection claims to explain the development of all species, starting from the most primitive organism to the most complex ones, like human beings. If so, then shouldn't the most primitive and simple ones have been wiped out from the Earth by now, so that it could be filled only with species which are superior and more complex?

Gertrude Himmelfarb gives the honeybee example in relation to this issue, when she relates how Darwin sings the praises of the honeybee for developing an excellent ability. To him, the process of natural selection made the bee's ability perfect, so this tiny being could get to the point where it can build the pores of a honeycomb by using a little bit of beeswax. Darwin was amazed by such architectural mastery, but he could not explain why and how other bees, such as bumblebees, which do not have the same talent as the honeybee, could still survive, even though they do not have any such special capabilities. The only thing that Darwin was able to say was this: "Nature left visible traces of its past handiwork on the way to perfecting its forms." However, this reasoning contradicts the idea of natural selection, which claims that the better model forces other relatives to disappear and it always wins. Yet even though they are less gifted than their relatives, so to speak, bumblebees are still able to grow up, reproduce and survive with their present physiological attributes. Contemplating all such plants and animals, along with the bumblebee, Himmelfarb asked the why there should be these living, not dead, remains, and why natural selection itself had not eliminated these imperfect and superseded forms.[10]

Evolutionists' answer to Himmelfarb's question was this: Bumblebees developed a strategy of survival in which they attacked honeybees and plundered their hives. Well, then these evolutionists also have to answer how have hundreds of very delicate, measured, and planned strategies of different types of bees developed coincidentally. They have to explain how queen bees, male drones, workers, etc., each with unique abilities, are selected in the social organization of bees.

This question has never been answered since those creatures, which should theoretically have disappeared by now through natural selection due to the existence of "superior generations" have actually not given up the race, and countless examples are still present everywhere. Thus, the fact that not only the more "gifted" creatures, but also the less so, survive renders evolutionary theory both incongruent and inconsistent. What is strange still is that Darwin saw natural selection as a slow process whereby each chosen, new feature would provide apparent benefits to the individual in the struggle for survival.

Natural selection can never be reconciled with the idea of irreducible complexity. As Michael Behe states, irreducible complexity is an important principle of structure and operation which is observed in living beings. In short, a system can function well and more productively with the presence of all of its parts. We may give Behe's mouse trap as an example: the absence of any one of the elements that is set up to catch the mouse—namely the arc, moving arms, caps, bait tray, and so on—will make the trap non-operative. In order for the mousetrap to catch the mouse, all of the required pieces have to be present in the appropriate arrangement or position. This principle, termed "irreducible complexity," is seen in all the organ systems of living beings.

Natural selection, functioning without intelligence or consciousness, cannot be used to explain the addition of each small piece, or function, of an organ or a body part in a precise way so that it might eventually be useful for the entire system. Stephen Jay Gould, of Harvard University, expresses this dilemma bluntly as follows: "What good is half a jaw, or half a wing?" Indeed, surely a half-organ or half-wing is not useful. In turn, N. Macbeth comments on the complete dependence of Darwin's theory on natural selection saying that Darwin's entire theory hinges on natural selection as a mindless process, as the impersonal operation of purely natural forces. If it is mindless, it cannot plan ahead; it cannot make sacrifices now to attain a distant goal, because it has no goals and no mind with which to conceive goals. Therefore, Macbeth thinks every change must be justified by its own immediate advantages, not as leading to some desirable end.[11]

In other words, every partial change would have to be somehow beneficial to the individual and to the species. However, if someone were to claim that there are millions of animals whose organs have still not been completed today, every reasonable person would reject this idea immediately. Yet this is what Darwinians are effectively saying. They cannot give any convincing answer to how natural selection gradually produces the "portions of all the pieces" which are necessary for an individual's survival. The organ which best manifests this impossible quandary is the eye. Again, Himmelfarb emphasizes a fundamental point about this in his *Darwin and the Darwinian Revolution*: Since the eye is obviously of no use at all except in its final, complete form, how could natural selection have functioned in those initial stages of its evolution when the variations had no possible survival value? For Himmelfarb, no single variation, indeed no single part, being of any use without every other, and natural selection presuming no knowledge of the ultimate end or purpose of the organ, the criterion of utility, or survival, would seem to be irrelevant.

An eye is indeed a marvelous and complex system. Among the parts, there is excellent synchrony which cannot be compared with anything else. For the veterinarian R. L. Wysong, two bony orbits must be "mutated" to house the globe of the eye. The bone must have appropriate holes (foramina) to allow the appropriate "mutated" blood vessels and nerves to feed the eye. The various layers of the globe, the fibrous capsule, the sclera and choroids must be formed, along with the inner light sensitive retina layer. The retina, containing the special rod and cone neurons, must be appropriately hooked up to the optic nerve which in turn must be appropriately hooked with the mutated sight center in the brain, which in turn must be appropriately hooked up with the brain stem (a grey matter in the center of the brain) and spinal cord for conscious awareness and lifesaving reflexes. Random rearrangements in DNA must also form the lens, vitreous humor, aqueous humor, iris, cili-

ary body, canal of Schlemm suspensory ligament, cornea, the lacri-
mal glands and ducts draining to the nose, the rectus and oblique
muscles for eye movement, the eyelids, lashes and eyebrows. All of
these newly mutated structures must be perfectly integrated and
balanced with all other systems and functioning near perfect before
the vision we depend upon would result.[12]

Indeed, that is the eye. Even Darwin admitted a couple of times
that he did not want to take the structure of the eye into consideration. He confided about this to his friend, Asa Gray, in 1860: "The
eye to this day gives me a cold shudder."[13] After all, we are supposed
to believe that each micro mutation, occurring gradually, contributed
some beneficial properties by selection while combining to form the
complex eye structure; and further, that those tiny changes, which
happened randomly or by chance, somehow resulted in such an
organ as the eye, having delicate sensitivity and a marvelous functioning, with neither a previous plan or final purpose "in mind." Thus, it
seems that even Darwin did not believe his own theory in relation to
this point, and this can be clearly understood from his own statements: "To suppose that the eye with all its inimitable contrivances
for adjusting the focus to different distances, for admitting different
amounts of light, and for the correction of spherical and chromatic
aberration, could have been formed by natural selection, seems, I
freely confess, absurd in the highest degree."

In fact, not just the eye but thousands of complex biological
systems can be given as examples that disprove the claim about
improvement through natural selection. For when they are analyzed more deeply, all types of systems which are present and available for use in living beings, are made up of components which can
only be used as complements to the whole, while the parts which
comprise the entire system do not, by themselves, have any beneficial function for either the individual's or the species' survival.

There are also other problems that do not seem to be resolved
and which profoundly shake the cogency of natural selection. For
instance, a long-term or planned idea cannot be congruous with

natural selection since each new intermediate form or feature has to be deemed useful or else be eliminated immediately. Effectively, Darwin saw natural selection as a kind of economic efficiency—as a way for nature to "increase productivity." Darwin surmised that in order to accomplish this goal, selection provided the necessary properties to the powerful so that they could surmount their rivals. In other words, those new features would be selected expressly so that an individual or species could resist or thrive in particular environmental conditions. According to Darwin, for a species to exhibit more new features than are needed could not be considered economical or natural, because he understood frugality, simplicity and moderation to be defining characteristics of nature. This means, however, that if an individual manifested features that were excessive at any moment in time but could be useful in future environmental conditions, then Darwin's entire theory of natural selection would be at risk since natural selection was indeed established on the idea of chance coordinating exactly with existing conditions. According to Darwin's scenario, there was no need for long-term planning, for competition was instantaneous—and a living being which adapted best to an existing environment would surely "win the race."

A person who studies evolutionary theory can be impressed at first glance in the sense that everything seems to be explained in a sensible manner and nothing appears to have been overlooked. However, if one looks at it fairly for a while and thinks about the subject deeply, one will easily see that the mechanism of natural selection has been interpreted with substantial exaggeration. Yet while the discrepancies and irregularities in the concept of natural selection have been evident for many years, advocates of theory of evolution have continued to strive to make the theory sufficient to quiete criticisms, which have increased day by day. Every time natural selection is criticized from a new perspective, its supporters simply reconstitute the fundamental principles of evolutionary theory.

In fact, it was not until the Nobel Prize winner, geneticist T. H. Morgan, expressed his opinion that the idea of natural selection was affected by faulty thinking and that no scientist had ever questioned what was being considered with all these arguments. About the definition of natural selection proposed by Neo-Darwinists Morgan stated that it may appear little more than a truism to state that the individuals that are the best adapted to survive have a better chance of surviving than those not so well adapted to survive."[14] Or, as remarked by Gertrude Himmelfarb, the survivors, having survived, are thence judged to be the fittest. [15] This determination surprised the scientific community since those words seemed to announce "the nakedness of the king," one might say. While Morgan directed attention toward a point which had seemingly never been considered until then, other critics have since further crystallized the error of natural selection which we are now faced with. For instance, as an important developmental biologist, C. H. Waddington has dealt the final blow, as it were, when it comes to displacing the theory from its effective designation as "a holy taboo." Waddington said for an animal being the most "talented" or "fittest" does not necessarily mean that it is certainly the strongest or healthiest or would win a beauty contest. Indeed, natural selection states that the fittest individuals in a population (defined as those who leave the most offspring) will leave the most offspring and it turns out on closer inspection to be a tautology (unnecessary repetition), a statement of an inevitable although previously unrecognized relation. Once the statement is made, its truth is apparent.[16]

Those that are able to survive after elimination by natural selection are necessarily the ones which will have reproductive opportunities, arising from their advantageous and favorable features; yet even they can only cause a horizontal change within their species by transferring the genetic potential given to them to their offspring. Let us take the example of the *Biston betularia*, or peppered moth, which is often used to defend the idea of evolution as a result of natural selection. Some individuals of the species are light-colored, almost white, while some of them are dark-colored. Before the industrial age, when there was no air pollution in Britain, the exterior walls of buildings were clean and light-colored; as a result, the birds that hunted the peppered moths were not easily able to recognize the white-colored ones, but they could readily detect the dark-colored ones. Thus, the number of dark-colored peppered moths was reduced, while the number of white-colored ones increased. It was not until the facades of buildings became dark due to industrial pollution that the dark-colored peppered moths secured some "camouflage" and the white-colored ones came to be more easily hunted; there was a consequent decrease in the number of white-colored individuals and a parallel increase in the number of dark-colored ones. Such a differentiation in the species, which was determined by the zoologist, H. Kettlewell, of Oxford University in 1924, is a horizontal change which does not represent a transition from one particular species to another—that is, it is not a vertical change.

As seen in this example, the peppered moths did not unveil a new feature that was not present in their genetic portfolio at the onset; instead, they showed a shift from the light color to the dark color within the limits of the color spectrum that already existed as part of their potential response to the environment. Since the lighter peppered moths were hunted easily when pollution pre-vailed, they simply died before they had an opportunity to repro-duce; meanwhile, the birds did not recognize the darker moths so easily, which allowed them to live longer and gave them a chance to reproduce. Further, according to Mendel's principles, the chance of producing the darker moths as offspring from the darker parents is higher, so the black-colored individuals became dominant in the surviving population.

Yet, whenever a person questions evolutionary theory, the advocates of the theory propose the miracle of the peppered moth (!) immediately as if it were absolute evidence for the existence of evolution. In biology books, pictures of peppered moths are given along with the impression that evolution has been proven for more than fifty years. But as mentioned above, the "peppered moth example" is actually and essentially a proof that demonstrates the *nonexistence of evolution* rather than its existence. But somehow, no one questions the mechanism of natural selection to such a serious extent. A change like that described in the example of the peppered moth—the peppered moth being white or black—can only be explained by the adjustment of the peppered moth to its environment within the limits of its genetic potential in order for it to maintain the continuation of the species. In other words, the light-colored peppered moth's apparent conversion to a black-colored moth is evidence for the preservation of the species, but not for the presence of evolution.

Natural selection, which is proposed as the fundamental mechanism and premise of the evolutionary hypothesis, is nothing but the biological principle of the food chain in the operation of the ecosystem which can clearly be seen as God's "Book of Nature," contrary to the perspective of evolutionists. Another point is that this principle is given as a necessary driving force in the natural operation of adaptations and thus for the incidence of horizontal changes by which strong, enduring individuals arise within the potential range of variability within a species—namely, "within the genetic limits of a species." In this way, the unlimited reproduction and distribution potential of each living species is balanced, and sick or unhealthy animals are effectively prevented from spreading and corrupting populations. In the meantime, food for innumerable living beings is provided. If the grey mullet us litter of five million eggs is considered, the importance of the subject matter is understood better. For if an offspring developed from each of the five million eggs, and all of them developed, each and every grey mullet us food supply would have to be taken into consideration. Yet, since all creatures do not have unlimited reproduction opportunities within the limited conditions on Earth, only a certain number of individuals from each species have the chance to live by means of this excellent balance. If we go back to the example of the grey mullet, about one million of the five million eggs will be food for other creatures, or they will simply perish and break up due to insufficient conditions while they are still in the embryonic stage. Another couple of million eggs will reach the larval stage, and then become food for small creatures, while others will reach infancy and then be food for bigger fishes. Thus, those which will actually reach maturity and have a chance at reproduction will be sufficient merely to provide the continuation of the next generation.

We have probably all watched television documentaries about lions and antelopes. If all lions and all antelopes were strong and healthy, lions would always run after antelopes to obtain their food, and the antelopes would always escape from them without being caught, so that the lions would just keep trying to catch them, and so on. However, both of these living beings need energy, a food supply, in order to be able to run in the first place. Indeed, since neither the lions nor the antelopes can stop running, according to this scenario, both of them should die of hunger and exhaustion. Nevertheless, such a dramatic example is never observed in nature because of the inherent variation within all species, such that some of the antelopes and some of the lions will be weak and powerless. The powerful lions will catch the weak antelopes and eat them, and then the meat left over from the antelopes will provide sustenance for thousands of other living organisms, such as hyenas, jackals, vultures, carrion crows, bugs, and bacteria. Meanwhile, the weaker lions will also die early since they cannot hunt to find food. By means of this food chain, a balanced population plan is achieved among predators and prey, and thus the ecological systems of the world are protected.

MUTATION: A MYSTERIOUS KEY, OR RANDOM BULLETS NOT MISSING THE TARGET?

The Cell and Genetics

Three common characteristics that living mechanisms bear are reproduction, variation and heredity. All morphological and physiological characteristics of a living being are transmitted by heredity from the male and female parents. In this way, every living being resembles its parents, but this resemblance is never an exact sameness; namely, even the offspring of the same parents are not identical except for monozygotic (single-egg) twins. While the occurrence of reproduction cells through meiosis is the most important feature in living beings having sexual reproduction, very rich variation still arises by means of the exchange of genes between homologous chromosomes. A gene or a part of a chromosome can cross over between homologous chromosomes—one coming from the organism's female parent and the other from the male parent—where the information about the characteristics of the same parts of a body are coded. None of the millions of sperm cells is identical to another; similarly, such richness in variety is also found in egg cells, which occur through meiosis, even though they are not as numerous as sperm cells. For this reason, an offspring formed by the fertilization of an egg cell by any one of the millions of sperm cells, each one of them having different properties, will be distinct from all other offspring. Not one of the billions of people on Earth looks exactly like another because of this mechanism. (The possibility of the exact resemblance of two people is actually one in seven trillion, unless they are monozygotic twins). Thus, even though every human being (assuming an absence of genetic or developmental deformities) has two eyes, two ears, one nose and two lips, the actual features of every individual human being's face form dissimilarly since there are countless possible variations in the chromosomal and molecular functions of genetic systems which, in turn, lead to infinitely many combinations.

Genes are huge molecules where the information about the structure, shape and functions of a living being are coded. They are composed of smaller molecules, and, in turn, those smaller molecules are composed of atoms, while those atoms are composed of minuscule particles. Even though huge DNA molecules and their composite particles, genes, are not alive, they are still the major material "cause" of a creature's becoming a living being.

The idea, first proposed by Lamarck, about acquired characters being transmitted to future generations through inheritance, was disproved by many researchers, and mainly by Weissman's experiment. Ever after, natural selection started to find acceptance as the driving force of the evolutionary mechanism; thus, how molecular functions were genetically programmed became a rich research subject among geneticists. Research in genetics also picked up speed due to advances in molecular biology and the fact that the discovery of Mendel's principles made it clear how living systems were coded and how such information could cause changes to occur during the reproduction process.

A living being's parents are not the same; nor are its offspring the same. Rather, along with their ancestors, the parents and their offspring get their genes (genotype) from the gene pool. While they all belong to that particular species, they also show their own characteristics (phenotype). Critically, however, the genetic mechanisms, mutations and recombinations—that is, the new arrangements of genetic information which are responsible for the same characteristics between homologous chromosomes through crossovers and replacements—which cause variations are independent of the actual needs of an organism. In other words, an organism's being in need of swimming does *not* cause a variation whereby hands and feet take the shape of flippers or fins. Thus, the occurrence of new varieties is completely beyond the will and knowledge of the living mechanism, and knowledge of those varieties is withheld until the living being is born. Indeed, each and every detail of these reproductive processes is known and accomplished by the Creator, Whose power and knowledge creates everything.

Both internal and external factors have an effect on the development of a living mechanism. External factors include ecological conditions, such as radiation in various intensities, temperature, moisture, and food. For example, the reason behind the difference between a queen bee and worker bee is a difference in their nutrition. Similarly, only after many years has it finally been understood that the temperature to which the eggs of a sea turtle or crocodile are exposed during their early development is one of the main determinants of the gender of the embryo; in fact, a turtle's or a crocodile's becoming female or male depends only on a few degrees' difference in temperature. Yet, this mechanism is not a simple phenomenon; rather, temperature merely plays a role in triggering critical biochemical reactions.

Internal factors include changes in the DNA molecule where the genetic program of the cell is coded, and such changes are known as "mutations." However, in order for any change that arises in the reproductive cells as a result of mutations which are precipitated by external factors to be transmitted to the offspring, these variations have to be transferred to the inheritance molecules (i.e., they have to be inheritable). This is because, as mentioned earlier, only variations which occur in the genetic molecule (that is, in the reproduction cells) can turn up as actual changes in phenotype in future generations; non-hereditary variations (i.e., modifications) simply cannot enact permanent or lasting changes that continue in future generations of a living being.

Even though the idea of "struggling to survive" is a valid biological principle, the idea of the "survival of the fittest" started to be seen as the main propellant force behind the evolutionary hypothesis, with its name, "natural selection," being inspired by the discovery of mutations. According to this perspective, in order for a species to be eliminated, either the environmental conditions had to change in extreme ways, or mutations, which appeared in the new generations of that species and disadvantaged them compared to other species in a given habitat, would have to dominate. But, critically, while mutations would make the species extinct, they could never transform it into a new species.

In that it is considered to be functioning on emerged mutations, natural selection can only happen when, or if, the changes in different parts of an organism co-occur—that is, every single gene which codes each crucial characteristic in which an alteration is required must effectively mutate with respect to a plan so that it can change at the very same time for a similar purpose. Yet those changes cannot occur by chance. For example, mutations observed in the emergence of subspecies (strains) are such changes, which are accomplished according to the Divine Will, and which simply belong to the creation plan which comprises the "genetic capacity" of the species.

A mutation is a permanent and transmissible change to the genotype (genetic material) of an organism that seems to occur suddenly at a particular instant in time. Mutations usually arise through physical or chemical external effects; they seldom emerge due to internal causes. Yet in order for a mutation to be observed as a change in a living being, the variation has to occur on the DNA (deoxyribonucleic acid) chain—specifically, on the gene which carries particular genetic information, and even more specifically, on the part where the information belonging to a specific protein is coded. This was shown for the first time in the laboratory during experiments on *Drosophila* (fruit flies).

All DNA has two chains which, as determined by the Will of Infinite Knowledge, are made up simply of sugar and phosphate groups. The two chains are composed of repeating sequences of four "bases": adenine (A) and guanine (G), which have a "purine" structure (being five- and six-membered "heterocyclic compounds") and thymine (T) and cytosine (C), which have a pyrimidine structure (being six-membered "rings"). Among those bases called "nucleotides," only specific pairs of them—specifically A with T, and G with C—can consistently undergo hydrogen bonding. Thus, the two helical chains, each coiled round the same axis, are held together and are described in terms of their sequences of "base pairs." However, in RNA—which is used in "translating" genetic information from DNA to create proteins—since uracil (U) is present instead of thymine (T), and RNA consists of only one helical chain rather than two, the genetic information which comprises RNA is "read" as consisting of three-letter "words," each termed a "codon," which are formed from sequences of three nucleotides in a row (e.g., GGA, ACU, AUU, CCU, UAA GUA). In sum, while the proteins which are created to constitute the fundamental structure of a living mechanism are composed of 20 amino acids, four different nitrous bases are created with the potential to form 64 possible codons of three nucleotides each. So, there is more than one possible codon which can encode most amino acids. In addition, some codons carry the specific information code that determines, or effectively marks, the beginning and the end of the protein synthesis (coding region).

Mutations can occur as a result of the exchange of any base with another base on a DNA chain, or by the insertion (addition) or deletion (removal) of one or more bases. The mutations that have occurred by the change of only one pair of bases on the genetic code of a DNA chain are called point mutations. In addition to this, there are also nonsense mutations, by which the new nucleotide changes a codon so that the codon no longer codes any amino acid, as well as missense (or nonsynonymous) mutations, by which the new nucleotide alters the codon so as to produce an altered amino acid. Mutations that emerge as a result of insertions or deletions cause more serious problems.

Since point mutations generally affect only one codon, they usually do not cause big changes. For example, a mutated codon can continue to code the same amino acid or another amino acid that does not change the function of the protein which can be synthesized. On the other hand, in some cases, even the change of a single nucleotide on a DNA molecule can cause drastic and deleterious results. For instance, the serious disease known as sickle-cell anemia arises due to this kind of point mutation. An offspring will have the disease if he or she inherits the mutated sickle-cell gene from each parent.

Further, in the event of adding or removing one or more DNA bases, big changes will occur in the structure of a gene. Insertion and deletion mutations cause "frame-shift mutations" that change the groupings of nucleotide bases into codons, so that there is a shift in the "reading frame," so to speak, during protein translation. For example, if we assume that a mutation has occurred on the first codon in the nucleotide sequence TAG GGC ATA ACG ATT, whereby an A base is added to it, the new sequence will become TAA GGG CAT AAC GAT T—and thus, information designating a totally different amino acid, or even a nonfunctional protein, will be encoded.

It must also be understood that the mutated DNA chains are paired, reproduced and transferred from generation to generation, just like normal DNA. A mutated genetic code can revert to its original normal form only by a new mutation. In such a case, the second mutation will serve to repair the original gene, so that it may regain its normal function. The effect of the first mutation thus sometimes disappears fully or partially due to the occurrence of a second mutation (called a "suppressor mutation"), even on a segment of the gene which is different from the first mutation.

Macro mutations that happen suddenly and cause big changes in the phenotype are not important for creating variety and change in a living mechanism since they do not let the living being stay alive. For instance, in the case of a zygote or a developing embryo which may be affected by radiation or a mutagenic chemical agent, it is possible for the living mechanism to have organ deficiencies, or severe physical deformities causing two heads or four arms, and so on, depending on the rate of change in the genetic program. Those who are born with such damage usually cannot stay alive long. In the case of chondrodystrophic dwarfism in humans, for instance, while the head and the body are normal, there are development anomalies on the arms and feet. This disease emerges due to a mutation on only one gene among thousands. Another chondrodystrophic anomaly is seen in dogs, by which they tend to be "long and low"; it is the result of a mutation that is seen to be disadvantageous for the dogs but advantageous for hunters, for such dogs can easily find hidden hollows such as rabbit holes.

On the other hand, micro-mutations cause small variations in the phenotype. Evolutionists interpret this genetic mechanism with an exaggerated claim, which goes beyond its scope, when they state that those micro-mutations will be deposited and diversify the species from generation to generation—that is, that the species will be converted into a totally different species. According to evolutionary thought, mutations can sometimes form a new organ suddenly after they have been deposited for a while. So they argue that there might be transitions from one species to the other through this mechanism. For example, they say that the gill of a fish can turn into the lung of a frog, or that the leg of a lizard can change into the wing of a bird. They also assert that the feet of a mammal that walks on Earth can transform into a fin, that the fat layer under its skin might get thinner through the shedding of its hair, and that the lactation mechanism, and even the process of birthing, might acquire a different character.

If we were to believe that the sudden results of micro-mutations are able to achieve gradual changes in species—though when the mutations happened, how each happened, and how strong an effect each had when it happened are all unknown—it would be necessary to accept that each and every mutation among innumerable mutations happens every time in the reproduction cells of the same individual among numerous populations as if each mutation were a conscious being with a purpose, which was aware of what it is doing, in that those mutations support each other, happen in a sequential order, and always reach their goal. For instance, in order for land mammals to be able to live in water, thousands of mutations that would cause hundreds of anatomical and physiological changes in their bodies would have to happen in the same animal's reproduction cells in a way that is controlled, occurring slowly in a certain order, with vital timing and direction. In addition, such changes could not have occurred in only one sex of a species; they would have had to happen both in the male and female at the same time—but such a case does not even have a place within probability calculations.

In fact, the claim that the occurrence of some small mutations in each living being will result in useful and advantageous characteristics for the living mechanism is not far from impossible. This is because even a mutation that alters a very small part of an organ causes a change limiting the operations of the organ and harming it. The boundaries of the occurrence of mutations are not that broad. Since they will damage the ideal structure of an organ, one or more mutations are disadvantageous for the organ. Besides, a change of an organ does not mean that the living being will completely change because such a case is harmful for that living mechanism and will cause its death (since the integrity of the organism's system is corrupted). For example, if the transformation of the gills of a fish that comes out from sea to land to lungs is accepted for a moment, since many changes are required—such as fins changing to feet, the disappearance of scales, the differentiation of the arches of the heart and aorta, the change of sense organs and the nervous system, and the adaptation of muscles to the walking position—and these cannot take place at the same time, then the simple transformation of a gill to a lung will not be useful enough, and it will cause the certain death of the animal. Similarly, any sound intellect cannot accept the viability of small coincidental changes in even tiny portions of an eye and brain, as these are very complex organs, or the viability of random alterations in the orderly encoding of the genetic program of an eye or brain as the result of changes in the nucleotide molecule that composes the DNA.

In effect, when mutations strike a perfect, orderly system that works harmoniously, harmful effects can be observed, and the disadvantageous results of mutations for a living being are thus well known. One may offer the following comparison: it is as possible for a reproductive animal to transform into a different reproductive animal by being exposed to destructive mutations as it is for a

1930s roadster exposed to a shower of bullets from a machine gun to transform into a new-model Mercedes.

Thousands of mutated cells come into existence in our bodies every day, and our immune system destroys 99.9 % of those defective cells, formed by harmful mutations, before they become risky for the body. But if the immune system is weak and not working properly, then the mutated cells produce cancerous tumors, which gain a fatal character as they reproduce themselves. In turn, if the mutations happen in our reproductive cells, these cause deficiencies which prevent ovulation or the production of healthy sperm—or which precipitate miscarriages in females, so that even if ovulation happens, the embryo is terminated at some point in the embryological development.

Despite this fact, evolutionists accepts that some genes are specially selected and are exposed to mutation one by one as a result of chance. In this case all of the three alternative consequences contradict reason.

First of all, it is generally accepted that it is unclear and uncertain how so many coincidences could co-occur. Also to accept any of the following three alternative scenarios—which are each revealed when sequential changes happen one after the other strictly according to chance—is absolutely contrary to rationality:

a. The first of these scenarios says, "Changes happened one by one in a mutated organ which improved the function of that organ." Yet such a case has never been observed in nature. A mutated organ is always observed to be imperfect and defective, since random operations which are executed on any system which exists in equilibrium only cause inequilibrium and imperfections to arise in such a system.

b. The second scenario claims sudden improvement in the flying mechanisms of birds as a result of small changes that happened in anatomical structure and physical processes. Here the problem of staying alive for a reptile whose body structure had changed variously in many aspects has been ignored, and the incongruity of "teaching" a reptile to fly has emerged.

c. The third scenario claims that the parts of a complex organ in different species developed coincidentally, and then those parts somehow gathered together and formed a cohesive organ. Yet, clearly, every piece of an organ like an eye, for instance, cannot assemble themselves and form an eye after developing from individual components in different living beings. Because a whole needs all the necessary pieces at the same time, a single component is not useful. When we multiply the probability of the occurrence of a beneficial mutation happening in one part of the "eventual eye" by the simultaneous occurrence of a beneficial mutation happening in the other parts of the "eventual eye," we face probability numbers approaching infinity which are obviously impossible. Similarly, the evolutionist assertion that in order to have a prokaryotic cell transform into a eukaryotic cell each independent organelle of a cell (the nucleus, centrosomes, golgi apparatus, mitochondria, chloroplasts, and so on) somehow goes into the prokaryotic cell and starts a symbiotic life whereby they begin to "form a complete unit" is doomed to remain imaginary. What is more, it is not possible for these organelles, each of which is equipped with a structure like a tiny factory, to emerge independently by chance. Even with highly developed modern molecular biology technology we have not yet managed to uncover their complete structure; thus arguing that their organic molecules have combined on their own to form these organelles is not valid either.

According to evolutionary theory, natural selection which operates through mutation causes a species to become extinct or to change vertically (transforming from species to species). However, any improvement that is in an early stage is useless unless it develops in a way that will properly function and will really operate. For example, let us assume that an element of a wing instead of a leg occurred by mutation in a reptile species. This is disadvantageous for that animal, and it is expected that the animal will be eliminated by natural selection due to the fact that it cannot perform its normal functions with a deficient, mixed leg-wing appendage. The fact that congenial evolution scenarios do not happen in real life is indeed very clear.

Let us make a confusing matter clear here. The above statements, on how mutations are observed one in a million times, and that 99.9 % of mutations are harmful, take the changes in genetic system (genome) into account. This is intended to explain the changes that would alter the organs and the systems of our body and that would happen in the genetic code as a result of adding new, beneficial functions to it. This should not be confused with the changes that take place in the cells of an immune system. The ability to cause genetic changes continuously is given to various lymphocytes in our immune system in order for the immune system to be able to fight against the changes in bacteria and viruses. In other words, as an essential attribute of their identity, bacteria and viruses are capable of frequent changes, which are made in their genetic systems, so that new varieties continually arise. Due to the emergence of such new bacteria and viruses, the ability of the host—for example, the human being—to stay alive depends on new abilities in the immune system which can cope with the attacks of these new strains. It is true that those changes which are observed in the cells of an immune system are actually mutations, in a manner of speaking. However, those useful mutations, which emerge to protect our lives, are not random; rather, they are encoded in the DNA which programs the operational principles of the immune system and the general operation of immune responses in the body. Moreover, such mutations are given for the protection of our lives, and they effect perfect—what some might even consider "miraculous"—changes that cannot occur coincidentally or by themselves to alter our species type. The expected mutations of evolutionists, then, are not the ones that lymphocytes conduct in their daily battles against bacteria and viruses, but rather the ones that occur in reproductive cells and which are somehow credited with transforming a gill into a lung, scales into hair, or a fin into a leg.

Experiments and Observations

Bacteria and the fruit fly are two elements most frequently used used in the experiments and observations that have been routine in the field to find out the level of changes mutations can cause. Bacteria are very convincing examples of the non-transformation of one species to another. They are the fastest reproducing elements of life. They constitute 75% of all living beings, and they have three million years of history—if their age has been determined correctly. They could cover the whole Earth, knee-deep, in thirty-six hours if they were not somehow kept under control. Bacteria also mutate much more than other living mechanisms; however, it has never been observed thus far that any bacterium has transformed into another living being.

The mutation rate of *Escherichia coli* bacteria, for instance, which mutate very frequently and have a division process which occurs about every twenty minutes, is between 10^{-5} and 10^{-10}. Yet, only the very same type of bacterium's more resistant strains have been successfully produced in hundreds of research attempts done on bacteria using various mutagens. Indeed, the main reason behind the challenges for the pharmaceutical companies in the genetic capacities of such bacteria types, which have gained resistance to many of the antibiotics available today, is those mutations. On the other hand, as mentioned above, a new bacterium type has never come into existence through such small and limited mutations; rather, only different strains of the same kinds of bacteria have ever been produced.

Yeasts, being single-celled living organisms, are found everywhere in our environment. They reproduce with very rapid division, producing alcohol and carbon dioxide, while metabolizing organic molecules. In turn, some yeasts can convert alcohol to vinegar because they have alcohol dehydrogenase, an enzyme that helps them to accomplish this process. This enzyme, being a protein, has a functional molecular component which consists of four subgroups that are loosely connected to each other. Each of those subgroups is made up of 347 amino acids. Due to such amino acids, the enzyme has a very high capacity for change and, critically, there is only one gene that encodes all of the subgroups of the enzyme. In other words, the subgroups are made according to the instructions on this one gene, and in this way the enzyme becomes functional. With the occurrence of just one mutation on this gene, the enzyme starts functioning deficiently. So, by means of just one mutation made in a laboratory, is it possible to create a scenario where the yeast cell can adapt without its enzyme function being damaged?

Yeasts can live without oxygen Yeast cells which are deprived of oxygen become dependent on the alcohol dehydrogenase enzyme. When a different alcohol compound, which synthesizes to a poisonous composite as a result of the action of the enzyme, was given to such deficient cells, the mutated yeasts showed resistance to such poisonous compounds. Studies showed that amino acids which were extracted from horses, and which are present at the same place in alcohol dehydrogenase, began to enter the yeast's protein. For this reason, the yeast enzyme started behaving like a horse enzyme; namely, it acquired resistance to alcohol. Those small changes are the same type of genetic event that can always be observed among members of the same species and that support the arrangement of the process of varieties and strains. The diversity that was caused by the molecular changes in different sections of the DNA chain representing the genetic material—such as the splitting of small pieces, shifting, folding and rejoining—are normal biological events that can always happen in all living cells. However, everyone can observe that yeasts cannot transform into horses by such processes, which are nonetheless described as "microevolution" by evolutionists. Because of this, it is more appropriate to use the notion of "micro-change" instead of "microevolution."

Grassé questions the matter by asking how the Darwinian mutational interpretation of evolution accounts for the fact that the species that have been the most stable—some of them for the last hundreds of millions of years—have mutated as much as the others do. Then, he answers that once one has noticed microvariations (on the one hand) and specific stability (on the other), it seems very difficult to conclude that the former (microvariation) comes into play in the evolutionary process. He says that the evidence forces us to deny any evolutionary value whatever to the mutations we observe in the existing fauna and flora.[17]

Being among the most experimented on of species, *Drosophila melanogaster* (the fruit fly) was prominent material for mutation experiments for many years due to its very short period of ovulation and development (12 days). In these experiments, X-rays were used to increase the insect's mutation rate by a factor of 15,000. By doing so, the reproductive frequency and environment that the species could have been expected to be exposed to over millions of years under normal conditions was provided; hence it was expected to evolve. But even though the mutation speed was increased that much, no living mechanism other than a simple "fruit fly," which admittedly underwent a few changes, could be achieved. It was observed that all the mutant organisms were disabled insects whose wings were not present, or whose feet became blunt, or whose backs became humpbacked, or whose eyes were not present. Not a single fly species having any superior ability whatsoever came into existence out of all those countless mutations.

Moreover, about the two experiments Ernst Mayr performed on fruit flies in 1948 he reports that in the first experiment, the fly was selected for a decrease in bristles and, in the second experiment, for an increase in bristles. Starting with a parent stock averaging 36 bristles, it was possible after thirty generations to lower the average to 25 bristles, but then the line became sterile and died out. In the second experiment, the average number of bristles was increased from 36 to 56; then sterility set in. Any drastic improvement under selection must seriously deplete the store of genetic variability. According to Mayr, the most frequent correlated response of one-sided selection is a drop in general fitness. This plagues virtually every breeding experiment.[18]

Macro mutations

After it was understood that transition from one species to another species (i.e., from a yeast cell to protozoa) was not possible by micro mutations, the emphasis was deliberately shifted to macro mutations, to see whether they are present or not. At the beginning of the twentieth century, Hugo de Vries (1848–1935) verified Mendel's principles once again through cross-pollination experiments on plants. Hugo de Vries, who observed the presence of different properties that were not seen in the wild samples and culture types of the *Oenothera lamarckiana* (evening primrose) plant termed such changes that suddenly arise in new generations "mutations" in 1886. Animals different from their parents had been known for centuries. As mentioned above, chondrodystrophic dwarfishness mutations that result in long-legged or short-legged subspecies, for instance, are established as a fact today. However, the transition of any dog into another carnivorous animal other than a dog has never been witnessed. Nonetheless, De Vries built up a new evolutionary theory using the results of his crossbreeding experiments. According to this theory, macro mutations were happening and natural selection had little effect on macro mutations. However, since even micro mutations are mostly harmful, and thus they are eliminated by natural selection, he should have given answers to questions about what kind of strange creatures macro mutations would cause, or whether they could survive or not. Also, considering that transitions from species to species were presumably possible by macro mutations, according to this idea, should we not of necessity come across hundreds of examples of those species in a state of transition from one to the other? Even more problematic, if everything actually worked like the flowers in the De Vries experiment, how many arms or heads would babies actually have, and would they be able stay alive? Yet, through it all, De Vries was insistent about the formation of all species as the result of strong mutations, which happened at the species level, according to his line of thinking, and he defended his theory of "mutations" to the last.

Today, the reality of changes in DNA which are called "mutations" is completely understood and it is accepted by almost everyone that De Vries's theory was exaggerated. Through advances in the field of genetics, it came to be understood that the appearance of those properties in the evening primrose that De Vries was working on arose as a result of chromosomal changes now called "translocations" and "deletions."

By crossing strains of a butterfly called *Lymantria dispar* from different geographical regions, German zoologist and geneticist, R.B. Goldschmidt (1878–1958) showed that distinctive properties are transferred to new generations and that the distinctive properties could be explained using Mendel's principles. However, he exaggerated this later on and argued that fish undergo a mutation whereby their chromosome numbers are doubled and they suddenly advance to become amphibians; then those amphibians transform into reptiles, and then to mammals, by huge jumps of macro mutations. Of course, geneticists found those claims unsupported, and thus they rejected those ideas. Chromosomes are very sensitive structures and playing around with them in such a way just reduces the chance of survival of the species.

It was acknowledged as a big disadvantage that not even a single new living species had emerged as a result of mutation experiments, and that on the contrary, mutations had been consistently observed to cause random and idle results—or that, instead of improvements to the living mechanism, they caused harmful and destructive regressions.

Yet until the 1960–1970s, the idea that evolution depended on mutation and natural selection was the general opinion of evolutionists due to the influence of the school of Thomas Hunt Morgan (1866–1945). Genetic recombinations did not enter their minds. Even though the occurrence of genetic crossovers between the chromosomes during meiosis was discovered in 1880, the pivotal role of crossovers in biological variation and diversification was neglected. Today, we know that the biggest source of the diversity within a species is the phenomenon of "genetic potential for production of new variants," which is otherwise called, "intrachromosomal recombination."

After all, it can be said that such mutations, being the cause of variations, each happen as a result of different genetic mechanisms, and they each have a function in maintaining the species equilibrium—but they do not cause or require any essential change. Thus, producing enough variation and diversity within the species guarantees the continued existence of the species. All genetic studies show that in the event that only one type of individual is grown for the sake of abating, or controlling, the variation in a species, the essential variability which is necessary for the continuation of that species is reduced after a while. Reproduction experiments also give results which are contrary to Darwin's arguments. For Darwin analyzed artificial production and then came to the conclusion that this causes the production of better animals and plants, which can survive more effectively. The biggest mistake Darwin made in this regard is that he confused "being more profitable" with "being more suitable or talented." By means of some techniques, a chicken that lays more eggs, a cow that gives more milk, a sheep that gives more wool, and a corn stalk that gives bigger corn can be produced. However, while doing this, the species' inherent ability to continue its own life, independently and longitudinally, is substantially reduced. This is because producers choose only the properties that seem profitable and ignore the other features of a species for economic reasons; in so doing, they actually harm that species. According to British geneticist, Douglas Scott Falconer (1913–2004) the improvements that have been made by selection in the domesticated breeds have clearly been accompanied by a reduction of fitness for life under natural conditions. Although this breeding under special conditions seems to be a success for purposes of profit, it is achieved at the expense of the overall ability of the species to survive. For the producers effectively deprive the species of its natural strength, which has been assigned to it, and damage its capacity to adapt; thus, they make it weak and less resistant to the harmful changes in its environment.

Grassé states that mutations look like a pendulum swinging back and forth, only within the changing capability of the genetic system, but they *never* cause evolution. They just make the present characteristics undergo some kind of change within a certain range around the central feature of the related characteristic.

Constituting one of the biggest deadlocks of the notion of evolution, some organs and behaviors termed "particularly equipped structures and specifically designed behaviors" are impossible to derive by mutation without a precursor, either suddenly or gradually. For instance, we cannot attribute any of the innumerable, marvelous and surprising phenomena which are witnessed in living beings—too voluminous to mention here—to organs developed by random mutations, or to randomly programmed neurons in the brains. We may consider just a few examples to make the point, such as a bat's radar, a dolphin's sonar, a firefly's glow, a glowworm's light, the bioluminescence of deep sea fishes, the silk-making abilities of silkworms, the honeybee's honey, the leech's ability to prevent blood from clotting, and the astounding migrations of everything from martins to storks, to conger eels, to salmon. Each could easily be the subject of an entire book on its own—and if we look at the subject matter from the point of view of "irreducible complexity," and we examine in detail the structure of those organs, we will be forced to admit that since all of the components of those intricate organs would not function if even one tiny part were missing, each of them was created as a unique and integral design—a work of art.

Can Struggle Explain Everything By Itself?

Darwinists see nature as a place of conflict where each organism struggles for its own benefit. According to them, natural selection guarantees the survival of those that have the most beneficial characteristics and the highest degree of efficiency in terms of their achievement. Even though the truth overturns this claim, such a presentation of nature has predominated so far. It is true that there is competition in nature, but this is not the only, or even the most dominant, feature of nature. By careful investigation of the behavior patterns which have been observed between animals for a century, it is now understood that there are many other forms of behavior which are present among animals, other than competition.

In their book, *Life: Outlines of General Biology,* John Arthur Thompson and Patrick G. Geddes point out the weakness of the claim that there is such a big struggle for life in nature. They argue that there is an exaggeration of part of the truth and underline the fact that while one organism intensifies competition, another increases parental care; one sharpens its weapons, but another makes experiment in mutual aid. For them the struggle for existence needs not be competitive at all; it is illustrated not only by ruthless self-assertiveness, but also by all the endeavors of parents for offspring, of mate for mate, of kin for kin. The world is not only the abode of the strong; it is also the home of the loving.[19]

In turn, in his book, *Algeny: A New Word, A New World*, Rifkin also emphasizes that natural selection looks good on paper, but as with so many theories, when exposed to the complex workings of the real world, the simplicity which made it so convincing in the first place turns out to be its undoing. Rifkin gives the example that proponents of natural selection would have us believe that there exists some neat casual relationship between a victim and a predator independent of their surroundings. For Rifkin one can almost visualize the entire contest taking place in an arena fenced off from the vagaries of the outside world. However, in the real world, the dexterity of the contestants often has little if anything to do with their survivability. It makes little difference whether one little ant's legs are more swift than another's or whether one chimpanzee is more intelligent than another when a fire or hurricane sweeps through a forest, killing, indiscriminately, everything in sight. Rifkin thinks natural cataclysms are responsible for a great deal of death and destruction, but the killing is so random and widespread that it is just a matter of pure luck which organisms are caught in the path and which are spared. It can hardly be said that those which survived and reproduced were in any sense of the word more fit; they were just more lucky.

Indeed, we should use notions of weakness and forcefulness while comparing individuals within each species kind. Some individuals of the animal species can be weaker or powerless, while others can be stronger. When all the members of a group belonging to the same species face unfavorable and difficult conditions, the weak, vulnerable ones die while the strong, resistant individuals survive. Yet, when a huge ocean wave hits the rocks, it kills everyone on it without considering whether an individual is weak or strong—and all could easily die at once, or disappear, in a major catastrophe such as an earthquake.

Some species actually have amazing defense and survival strategies, distinctive attitudes given to them as divine urgings (what evolutionists call "instinct"). Some bison types, for instance, gather together in a circle against a ferocious animal like a lion. They stand in such a position that their horns face outwards, and their hind quarters are inside the circle; in this way, they can resist attacks while protecting their vulnerable offspring, who are sheltered in the middle of the circle. Such an attitude makes even a lonely, weak ox very strong through collective behavior.

In addition, the attitude of self-sacrifice to preserve its offspring (for the continuation of the next generation) is also seen in some other species. Such altruistic behavior is not only beneficial for the individual but also for the group. Yet while the group's total productivity increases, the altruistic individual's own productivity might actually decrease. In other words, while "group selection" supports altruism and leads to the viability or extinction of an entire group, the organism's own selection supports selfishness and allows the reproduction or the death only of the individual. So, can selection steer, or improve, such sacrificial behavior in favor of the group or the individual? Of course, right after this question, the following comes to mind: "Is there any purpose behind the concept of selection?" For if there is, then the presence of the Owner of such knowledge and power, Who makes selection purposeful, will be sought. The answer to this is that since such a perfect and conscious mechanism cannot proceed by itself or coincidentally, the existence of a Creator is deemed both certain and absolute.

Another crucial point we should emphasize here is that camouflage and mimicry (a deceiving characteristic which serves as the chief means of protection of the weak against powerful predators), both being means of survival other than struggle as well as common living behaviors, are adjusted very critically.

As highlighted by Bergson in his book, *Creative Evolution*, it is inevitable that we will err in ascribing knowledge and will-power to the notions of adaptation and selection if we do not attribute the excellent behaviors which are observable, and sometimes called "instinct," to infinite knowledge and power, and if we do not see those attitudes as "divine urging."

ADAPTATION, OR GENETIC INSURANCE OF SURVIVAL?

Being a good observer, Darwin recognized the rich variety which can be seen in the animal world. However, not knowing the genetic mechanism behind these varieties was deceptive for him. Noticing the small changes within species, Darwin arrived at the conclusion (by way of a "short cut") that these could create a transition from one species to the other. That idea had an exciting and appealing side for everyone. Critically, though, the fossil record and modern techniques in animal breeding display the fundamental conceptual errors of both Darwin and his current advocates. Indeed, variations within a species that are harmonious with the environment simply increase the species' ability to protect itself, and "insure" future generations against serious environmental changes; in effect, those changes ensure the maintenance of the "biological borders" which that species has had since its creation.

It can be said, then, that variations are not vertical—rather, they are horizontal. In other words, the genetic combinations that arise as a result of meiosis or of changes in the genetic code as a result of other mechanisms cause diversity and richness within any species, but they do *not* give any opportunity for transition from one species to another new species. The multiplicity of variations within the species is a kind of insurance for the continuation of that species' generation. In this way, the existence of the species is maintained despite the difficulty of surviving under different environmental conditions. The key factors which will make it easier for any species to continue in future generations will be how much it can reproduce, and the extent to which its offspring sustain this portfolio of genetic variability. Even though some of its offspring are certain to die in extreme conditions that may arise suddenly, some of them (those which are more resistant to difficult conditions with respect to their genetic code or genetical potential) will still have the chance to survive, and the continuation of the species will be ensured through those individuals.

In the statement, "Adaptation is a result of selection, such that desert plants achieve survival by adapting themselves to dry weather conditions," the term "survival" (selection) expresses a result. Yet when the notion of the "survival of the fittest" is used along with the concepts of selection and adaptation, there arises a "barren cycle"—a circular definition, as it were. The answer to the question, "Which of them survive?" is given as, "The fittest ones"; and the answer to the question, "Which are the fittest ones?" is then given as, "The survivors." Thus, we are faced with an absurd and tautological statement which can be expressed succinctly as, "Those that live are the survivors." Returning to the example, in order for desert plants to adjust themselves to conditions, they first have to undergo selection and then adaptation; thus, after the elimination of unfavorable individuals, the rest of the population is said to have "adapted."

According to the reasoning of evolutionists, if the fittest ones are those that can adapt (i.e., those undergo selection first), then in order for them to emerge, there has to be a process of adaptation (i.e., those which undergo adaptation first). Such a circular definition is essentially a paradox which can only be resolved by considering it as a mechanism placed by the Creator as a central feature of nature, and by refusing to attribute "willpower" to natural selection. On the other hand, should natural selection be accepted as an "authority" having willpower, foresight, and knowledge—while also denying any "cause" which created the balanced operation of the ecosystem—then it is not possible to resolve this paradox. The degree of fit of an organism's survival is determined by its strength in life (health, fitness, power) and rate of reproduction in certain environments and populations. However, such an achievement does not only depend on the deterministic mechanisms of biology. The survival of the weak along with the strong can only be explained by such concepts as cooperation, solidarity, compassion and sacrifice among animals—thus, there is the implicit necessity in any analysis of taking the whole animal population into account.

Adaptation is a concept which fundamentally expresses the suitability of some individuals' genetic substrate for their survival as a response by which they can adjust to various environmental conditions, but it does not have, and is not, a self—a "being" on its own. This point should never be ignored when accepting adaptation as a causal biological mechanism limited by genetic substructures which were put in place by Divine Will in order for the ecosystem to operate in an orderly and harmonious fashion to ensure the continuation of the species.

The process whereby an individual's physiology and phenotype encounters and responds to environmental conditions is called "physiological acclimatization." The increase in the number of red blood cells among those who climb high mountains can be cited as a common example of such a phenomenon. On the other hand, the same physiological event is described by evolutionary theory as a process which has emerged by the special forces of natural selection, which increases the suitability of a living being in regard to its environment and changes the species gradually. Indeed, as a response to the environment, the individuals of a species might change—but to what extent?

For instance, Eskimos have a high-fat diet to be able to live in the North Pole; however, they do not suffer from the heart disease and some of the cancer types that are generated as a result of high fat in the diet. This is because there has been a favorable differentiation in Eskimos' physiological process to metabolize the fat compatible to the polar climate. This differentiation, however, has not transformed Eskimos to a species other than human; it has only remained at the level of subspecies (race).

Adaptation, being an individual characteristic of organisms, causes an average increase in the population's suitability to the environment—but it does not require or entail an increase in the growth rate of the population. In order to understand adaptation as a measure of survival, as well as the capacity for reproduction of a genotype with respect to other genotypes, we may make use of the comparison between the shape of a structure and its design by an engineer who has created it for a certain purpose. For instance, the design of a plant as an aphyllous (leaf-less) or spiny one, living in a very dry environment like a desert, is vital for its survival in that it ensures optimal water retention under difficult conditions. Simply put, any other type of biological leaf structure would not allow it to stay alive. The exquisite plan belonging to such a plant, called its "genetic code," obviously demonstrates that there is an Infinite Power Who designs that structure. Other examples of such intricate designs include the butterfly's harmonious colors, which are ideal for their diverse environments, and the camouflage of insects, by which they are protected from their enemies as a defensive adaptation. Yet simply naming a biological law does not necessitate disregarding the deep wisdom, love, mercy and compassion underlying these phenomena and the Artist, the Owner of infinite power, Who designs all of creation—in other words, asserting biological realities does not entail disrespecting or ignoring the Almighty. Saying, "Living beings have coincidentally developed characteristics which allow them the best fit all by themselves in order to adjust, as a species, to the environment," is nothing but a very shameful statement and a deliberate effort to cover up the truth.

Different animals that live in the same environment and within a common region do not have the same behaviors, that is, they do not respond in the same way in adjusting to the same environment. For instance, observing how a wild female bee digs a hole to store a dying grasshopper as food for her babies, one might ask why other species in the same environment do not exhibit the same behavior. Rather, the presence of specific activity patterns for each species is readily observable. In that case, one has to jointly consider the particular behavior patterns of each species and its manner of adapting to the environment. While studying the subject matter of adaptation, it is very important to compare different species through experiment and observation. As a case in point, in order to see if their body structure is favorable for swimming, the following may be understood from the analysis of the hydrodynamic features of sharks. It is accepted that the body structure and sense of smell of sharks are adaptations which directly affect their ability to survive by allowing them to seek prey and hunt in the water, and to swim quickly so they may escape from their enemies; seen from this perspective, a hammerhead shark's shape could be seen as being contradictory to such hydrodynamic advantages. Further, there are a great many creatures demonstrating wildly different appearances and physiology in the sea; none of them looks like the shark, but they still continue their lives in the most ideal conditions for themselves. Thus, it is clearly seen that all species are created with sufficient genetic potential to equip them with unique and particular recompenses and mechanisms to ensure their competitive advantage. If the general shape of the species is not very hydrodynamic, then another feature will compensate—be it the number of its fins, its ability to be camouflaged in the coral reefs where it lives, the protective property of its skin or its poison, its celerity or agility, and so on—something will serve to optimally offset what might otherwise appear to be its deficiency.

From the examples above, we deduce that not every feature which is considered in the domain of adaptation is acquired subsequently; rather, most of them are given to the species from its creation. As a matter of fact, if it were otherwise, the species could not stay alive long. It is also clear that there are limiting factors on adaptation. If having legs is an advantage, then when snakes compete with lizards in the same environment, snakes should always lose the competition—and yet they do not. Conversely, if being without legs is the advantageous condition, how could the presence of both legged and non-legged lizard types be explained? The fact that all snakes have no legs shows that the particular structure and design which belongs to a certain form was not given to every group of living beings; that is, there is a restriction applying to every type in terms of different aspects. Along with the presence of these "restricting factors," the living being is allowed to change within the limits of that species starting from embryonic development. Only when there are extreme genetic changes that strain the limits of the species does "evolution" occur in a complex organism—but when it does, it results in miscarriages and deformities, which are not viable and do not stay alive as they exceed the bounds of this restriction.

Certain sections of the genetic system are very fixed and unchanging. Since these are related to the vital characteristics belonging to a taxonomic class of living beings, we put the animals in big categories—such as fish or birds, carnivores or herbivores, or tortoises and snakes— accordingly. It is also possible that in the living being's genotype, there might be restricting factors restraining the occurrence of mutations which would change the basic features of the class in which it was created. Therefore, for instance, we can easily distinguish birds, reptiles, worms, and insects.

In order for the species to display itself in various phenotypes through new variations, the reality of the restricted factors that limit the changes occurring in the genotype are not known yet. However, as is known from reproduction phenomena in nature, while these provide the protection of the species' own original characteristics by means of an excellent restriction mechanism, the small changes that cause richness and subspecies are not obstructed. For example, humans are able to live in both polar zones, where the temperature is around - 60 °C, or in the Sahara, where the temperature is about + 60 °C; similarly, they can live in forests, mountains, tropics, lowlands, and so on—and they may even undergo some physiological changes while adjusting to such diverse geographic and climate ranges.

When we come across people from the different geographical regions of the world, some differences in skull shape, cheekbones and nasal bones, forehead projection, the width of the face and the shoulders, and the height, color and proportion of their body parts all give us a hint about which part of the world they are from. Nevertheless, these characteristics emerging within the range of available genetic variability do not alter the human species into another type of species and do not change the crucial characteristics which define a human being.

Here, a change in this regard could certainly be verified. Environmental conditions can sometimes change very radically, but since the genetic capacity of a living being cannot respond appositely to such new circumstances, it is possible to witness the death of that species and the extinction of its generation. For instance, the extinction of dinosaurs can be given as an example. According to the information we have today, they lived in past geological ages, but since they were not designed to have the capacity to adapt to the disaster which happened 65 million years ago, dinosaurs disappeared. Nonetheless, there is not a single shred of evidence showing that dinosaurs became smaller and transformed into modern-day lizards.

In fact, there is no such necessity, since all the characteristics of organisms have to be adaptive and well-adjusted. The important thing is the adaptation of vital properties. Rather than the consistency of all features considered one by one, both the integration of these features during the development period, and the variability accorded by the pleiotropic effect of the genes (whereby one gene is responsible for, or affects, more than one phenotypic characteristic), are essential. Further, not all characteristics are programmed genetically; some are especially coded in a way so as to emerge under the influence of the environment or learning. The ability of some genes to display the coded information that is present in their true nature to various degrees, as well as situations in which a protein is not able to be synthesized by an existing gene, all show the "open aspects" of the genes in regard to environmental effects. Some of the behaviors of humans and animals can be learned, and cultural heredity is also possible. It is clearly seen from observations of nature that the power of adaptation is limited within species groups. It should also be clearly understood that a harmonious system, integrated with the ecological conditions, is placed in the genetic codes of living beings for continuation throughout future generations.

Natural Selection and Adaptation from the Perspective of Creation

Selection can be explained as "the general name for all types of processes related to the survival of individuals which accomplish the struggle of life." Since the distinction between genotype and phenotype was not known in Darwin's time, living beings were thought to have simpler systems of inheritance that could be changed easily, as compared to the reality of the complex mechanisms and processes which are understood today. A great many conclusions about inheritance have had to be changed in recent years as researchers have become aware of the marvels of genetics and of the inherently miraculous molecular design of the coding which defines this system. The biological world is classified into systems within one another, including progressively larger components as one moves from gene, chromosome, genome, organ, organism, species, genus, family, and to group. Thus, the answer to the question, "At what system level is natural selection (supposed to be) working?" becomes a critically important one if one hopes to understand an organism's traits.

For a long time, rather than considering the gene or genome, an individual living being has generally been accepted as the selection unit—that is, the unit of operation on which selection is presumed to be working. However, the functional variation in the DNA molecule which is now known makes the analysis of the genotype from a reductive or atomistic point of view null; on the contrary, this knowledge necessitates that countless components and systems be studied using a holistic approach. A gene, having its own molecular existence, is both stable and inheritable, but it is not an independent structure. Cells and organisms carry the genes; in this regard, they can be thought of as "containers" for the genes.

According to evolutionists, changes in gene frequencies within a population cause selection completely dependent on coincidence, or cause "genetic drift," a specific variation in the genes of a small group. In effect, genetic drift is a statistical effect which occurs within groups of the same species wherein there is a small gene pool, and it is dependent on some natural processes which are an inseparable part of the general equilibrium. In this way, it causes some genetic traits of a small group belonging to a particular species to disappear, become "shielded or hidden," or even to become more common—all of this being independent of the reproduction rate. Whereas certain alleles (variants of a gene) are carried by many individuals in bigger populations, so that the balance in a gene pool does not normally change, genetic drift permits unfavorable biological conditions to emerge. In this case, an important factor which is called "founder's effect" also arises; this concept is based on the fact that some individuals within a migrating group which is separated from the larger population would have different alleles represented than the main group. So, the first founders of the migrating group would not be truly representative of the main, or entire, population. For instance, let us assume that within a community, some people have blue eyes while others have brown eyes. If only the people with blue eyes migrate to a remote place due to strains on land use, for instance, and establish a new community there, all of the children born in the new community will have blue eyes—and in regard to this trait, they will be differentiated from the people of the previous community.

However, the changes which occur as a result of genetic drift never form a new species; rather, they are simply changes which diversify the species' present capacity in various ways. In other words, genetic drift has the capacity to increase richness and variation within a species, but it does not add any new features to the genetic code.

DNA is open to functional variations, which ensures that species can adjust to varying environmental conditions. Degrading such an excellent system to suit a reductionist or atomistic viewpoint means seriously understating this amazing phenomenon. That is why most geneticists working today accept the genotype as a holistic and multi-component system. Further, the "selection value" of any particular gene is understood to depend on the structure of the genotype—in other words, all of the genes—with which it belongs.

It is also important to comprehend that most of the changes which occur in gene frequencies are neutral and do not have any selective importance. The examination of those types of change clearly shows that there is no possible way in which mutations might cause the evolution of a genotype by means of natural selection.

Indeed, being lower and mostly neutral, the selective value of the changes which occur at the molecular level are crucial in terms of protecting the originality of the species. Otherwise, the concept of "species" would become vague, and the genotype would be nothing but a "gene soup" that could transform into anything. For this reason, it can actually be said that natural selection is a mechanism given for the protection of the generation of the species by means of optimization, stabilization, cleaning, organizing, and ordering.

Further, even the flexibility of physiological adaptations that are not strictly genetic in origin is under the control of the genes.

Important elements and functions of the genetic code—such as organizing genes, moving particles, and repetitive DNA sequences—were made clear by the discovery in 1966 of "polymorphism" (numerous variations which allow quite different types to exist in the same population of the same species, such as ants or bees) in the genes which code the synthesis of enzymes. Thus, the necessary potential—the full complement of abilities—which is necessary for a living being to survive has been put into its biological structure wisely by the Owner of Eternal Power and Knowledge as a program which integrates its composite and complex elements with environmental conditions.

There is no possible way to predict the biological events that reproduction cells and zygotes may face when a new living being is born, but if we consider the possible circumstances only from the biological point of view (not taking divine wisdom into account) they can be described in brief as follows:

a- the loci (the specific sites of a particular gene) where mutations can occur on the chromosomes;

b- the loci where chiasms may occur resulting in crossing over;

c- the splitting of the chromosomes;

d- which specific reproductive cells among billions of them actually live;

e- which specific sperm and egg cell is then chosen, and why;

f- the development process which results from the combination of characteristics of the fertilized egg, and the influences of the outside environment—none of which can be predicted.

Furthermore, the phenomenon of "pleiotropy" (whereby a single gene affects more than one phenotypic trait) is also evidence for destiny in reproduction because, for a gene to be "readable" in more than one form shows that selection is actually a "hidden" and probabilistic event in terms of destiny.

At the time it was first proposed, natural selection was not accepted. The main reason behind this was Darwin's lack of comprehension of the importance of variation and his lack of supportive examples from nature. Due to the fact that, on the one hand, a probabilistic explanation could hardly have been understood during a historical period when determinism was so popular, and, on the other hand, all biologists used to believe the typological thought (essentialism) which had been influencing the West since Plato—the presence of constant, stationary, and invariant forms had been accepted. According to this line of thinking, there was only continuously changing "embroidery" on such stationary forms. Furthermore, in regard to the dissimilarity between the two individuals involved in sexual reproduction, it was not known that the two critical cells are different depending on the various activities of organizing genes. For this reason, the view that each living mechanism has a unique and special structure, open to change, was seen as an assault on the belief in creation. However, as a reality of the Creator's handiwork, the variations in populations actually signify the emergence of individuals with unique structures that belong only to them. In fact, this *is* the richness of creation. Furthermore, since the notion of "population" had not yet been advanced at that period in time, biologists used to understand events on an individual basis. Later on, when events came to be considered at the level of population, it was understood that average values were just an abstract number, so to speak, so that the acceptance of natural selection became easier. Yet, even at that early point in the discussion, a particular orientation prevailed over the interpretations of natural selection, which effectively directed both attitudes and conclusions falsely towards atheism.

It is obvious that natural selection exists as an intrinsic aspect of the food chain among living mechanisms, and it works with adaptation as an assurance for the protection of the generation. However, as the correlation between any two events cannot be evidence for "causality" –for a cause-effect relationship between them—comments made about an event do not entail that both the reason and the result are explained. Most of all, this is clearly seen when notions such as the survival of the fittest, adaptation, and sex are factored into the biological system. Developing mathematical models which take the gene as the primary unit of selection, population geneticists have tried to explained selection at the level of the gene, but as a result of neglecting to consider the whole individual, or the whole organism, those studies have yielded false results.

As to the primary view of modern evolutionary theory, evolution is only a process of adapting to suitable environmental conditions, or of taking advantage of opportunities that arise as a result of environmental changes. Since it does not have a definite purpose, the particular way in which it develops cannot be predicted. Should such point of view be accepted, the natural conclusion which one would reach is that everything—nature, humanity, and the human body, including its complex anatomy and physiology—is the fruit of coincidence, and that everything has arisen by itself from chaos.

In turn, the fact that all living beings having a common system of genetic coding in terms of basic molecules is presented as evidence that they have a common ancestor and origin. However, very same phenomenon is, in fact, the seal of the Creator's unity and powerful evidence of how He creates countless varieties using the same material.

FUGITIVES FROM THE GENE POOL

Interpreted as a mechanism of evolution, "isolation" is a phenomenon that may actually be applied to the past, too. According to evolutionary thought, a population consisting of individuals belonging to the same species might have been divided into many subpopulations for a variety of reasons. For instance, a population belonging to species A can be divided into a number of new populations, A1, A2, A3, A4, and so on, due to migrations or different geographical factors. If those new populations cannot come into contact with the original population in any way, so that they become entirely isolated from it, then they will only have a chance to reproduce among themselves, that is, they will only have the opportunity to exchange genes in a more limited gene pool. Thus, each small population will become a new gene pool by itself, and due to this phenomenon of isolation, it will not be possible to add new genes to that pool. As a result, the group of individuals constituting this gene pool will only be able to transfer the genes which are presently in the pool to each other. In this way, certain traits will start becoming dominant in each gene pool in time. As this isolation continues over many years, each dominant trait in the gene pool will become even more obvious, and eventually, it will be evident that this population, which separated from the original group thousands of years ago, is comprised of individuals who are substantially different from those in the original population.

According to advocates of evolutionary theory, individuals belonging to the new gene pool become so different from the former, ancestral population that they can no longer be paired with individuals from the original gene pool and they cannot produce new offspring since they have become, effectively, a new species. According to the well-known systematician Mayr, a species is "a group of actually or potentially interbreeding populations that are reproductively isolated from other such groups," that is, one type of species cannot interbreed with another species naturally to yield fertile offspring.

As briefly explained above, the differentiation of individuals in dissimilar gene pools through the mechanism of isolation is true; however, by exaggerating this phenomenon of differentiation evolutionists propose the claim—which is actually impossible to verify, experience or observe—that new species are created. However, in order for such an assertion to be confirmed scientifically, very long-term studies, in the order of millions of years, would be required. Thus, the existence of those mechanisms, which are necessary elements in the fabrication and pretense of the evolutionary process—and which, critically, cannot be falsified, and thus do not meet the most basic criteria required of scientific theory—must be considered to be a strictly non-scientific claim, a mere assumption.

Ultimately, then, while they do belong to unique gene pools, individuals who have become different from each other over a long period of time through isolation do not comprise an entirely new species; rather, they are just the subtypes of the same species. When the isolation among those subtypes is reduced or eliminated, the individuals of both populations can successfully interbreed with each other and produce crossbred strains.

As a matter of fact, this was achieved in the laboratory with interbreeding experiments between different subtypes, and crossbred offspring were reliably obtained. As we are a type of living species, the same phenomenon occurs among human beings. After the individuals of the first population of humans initially reproduced, they began to spread to different regions of the world. Since they were now far away from their parent population and completely isolated from it, they formed closed gene pools through marriage only among people from their own subgroups. Over time, as some genes became dominant over others (for example, as reflected in the darkening or lightening of the skin color; eyes becoming either more slanted or straightened; or hair acquiring or losing its wave or curl; and so on) due to the environmental conditions and selective interbreeding, groups with some prominent and distinctive features resulted. However, these groups certainly do not each comprise independent species—rather, they are simply distinct strains of the human species. Human beings from all groups can intermarry to produce offspring who are commonly witnessed in the world today as children of mixed heritage.

As briefly mentioned above, the two major mechanisms seen in the process of the emergence of subspecies by means of isolation are as follows:

a. *Geographical isolation:* This type of isolation occurs when part of a population of a species becomes geographically isolated from the remainder as a result of such geographical barriers as mountains, rivers, lakes, canyons, deep valleys, and so on. For example, a tailed salamander called *Mertensiella luschani*, which lives in the western part of the Toros Mountains in southwestern Turkey, has about eight subspecies, as determined by taxonomists. Those subspecies, which have been separated from each other by certain mountain ranges and valleys over a long period of time, are very slow-moving animals which do not have capacity to migrate in order to remove the isolation barrier; therefore, each of the subspecies has become different from the rest of the population in terms of color and markings.

b. *Ecological isolation:* This generally follows geographical isolation. As is commonly known, in different geographical regions, ecological conditions generally vary, too. If one of the individuals of the same species were living in a forest, one were living in a steppe region, and one were living in high mountains, for instance—so that each one was specifically adapted to its surroundings and did not migrate to other regions—they could not come together to interbreed, even if there were no geographical barrier between them. Consequently, since each of them would interbreed only within the gene pools that they form under these diverse ecological conditions, after some time, a new subspecies, with predominant genes which dispose individuals to that habitat, would be produced.

Other than these two, evolutionists differentiate three more types of isolation: genetic isolation, temporal isolation and reproductive isolation (either gamete or zygote-based). This approach is based on the claim that population which are separated from each other "take a form which cannot interbreed with other populations" after some time. According to the claims of evolutionists, the populations, which are initially capable of interbreeding, eventually attain very distinct characteristics upon their lengthy separation from each other—as a result of chromosomal changes resulting from gene mutations—and thus, these two different populations will not be able to reproduce when they interbreed since their gene series will not be compatible (genetic isolation). In the case of temporal isolation, those distinct populations start functioning in different seasons, so

they simply cannot find each other to interbreed. In the case of reproductive isolation, either the structure of the reproduction organs of
individuals in different populations, or their reproductive behaviors,
change through mutation so that they cannot reproduce even if they
find each other. All in all, for advocates of evolutionary theory, those
three isolation mechanisms also result in new species.

In fact, however, these claims about evolution can never be
proven, whether by experiment or by observation. As such, it does
not seem possible for the reproductive organs or the genetic codes
of thousands of individuals belonging to the entire population to be
changed by random mutations without ruining the species' normal
structure. In other words, even though such a change might occur
in one single individual, that extreme change would not have any
importance in the whole population since that mutated individual
would actually die and disappear after some time. It is certainly a
weak assumption to claim that while all of the physiological properties of various subgroups separated from the same population pre-
dispose it to function in the same season, as a result of a mutation
a need to be active during different seasons will emerge in all of the
individuals. The case where a living being, active in winter, has
become inactive by mutating has never yet been observed, and,
further, the formation of subspecies can typically be witnessed by
examining the fauna of islands.

Living in the Galapagos Islands, a species of finches which
became popularly known as "Darwin's finches" and which have occupied books about evolution, are first and foremost the most speculative materials used by evolutionists.

The Galapagos Islands, which are made up of thirteen main
volcanic islands, are about a thousand kilometers west of South
America. They are distributed around the equator, and the biggest
one is 112 km long and at most 32 km wide. The surface area of
some of those islands is not more than a couple of square kilometers.
And most of them are closer to each other than 100 kilometers.

Even though the physical properties are not very attractive, Darwin found those small islands striking and worthy of his attention. These islands are the only habitat for many animal and plant species living there. Darwin states in his voyage notes that there are at least one hundred native floral plants, dozens of unusual insects, and about thirty unique bird species. Besides, there is also a giant turtle species which is peculiar to this area, and two similar lizard species—one which lives on land, and the other which lives in the water. Among those, the marine iguana is vegetarian, feeding only on seaweed; its limbs are held to the side and its dive is typically shallow, while it can stay submerged for considerable lengths of time. The most remarkable aspect of these animal groups is that most of them—turtles, iguanas, finches, and others—are different from one island to the other in such a way that the forms special to each island *look* as though they belong to distinct species. Of course, the fact that subspecies could arise over time due to reproduction only among themselves, through the effects of genetic variations and the isolation of gene pools, was not known at that time. And yet it was specifically those apparent variations among the finches which triggered the "first structures" of evolutionary thought in Darwin's mind. He points out the following in his notes:

> The distribution of the tenants of this archipelago would not be nearly so wonderful, if, for instance, one island had a mocking-thrush, and a second island some other quite distinct genus; if one island had its genus of lizard, and a second island another distinct genus, or none whatever – or if the different islands were inhabited, not by representative species of the same genera of plants, but by totally different genera.... But it is the circumstance, that several of the islands possess their own species of the tortoise, mocking-thrush, finches, and numerous plants, these species having the same general habits, occupying analogous situations, and obviously filling the same place in the natural economy of this archipelago, that strikes me with wonder.[20]

He simply could not acknowledge the creation of bio-diversity, composed of completely different species or subspecies, among such proximal islands. As the fixity of species doctrinal approach assumes immutability, Darwin's sentiments revealed that he could not comprehend the way in which different species could be created one by one, specifically for those small islands, where some of the environments consisted only of a few sharp-pointed rocks. The only possibility for the creation of those variant species which he could fathom, or which he could allow, was the idea that somewhat similar species living on different islands had arisen from a common parent species through evolution. In fact, however, his misunderstanding may be traced to the fact that he personally lacked a profound belief in a Creator God Who is Almighty and All-Knowing, and he did not have comprehensive knowledge of God. For the divergent species which he observed could actually be explained as representing variation within the species rather than any sort of transition from one species to another. Such diversity could also be explained as being the separate creation of distinct species on a parent continent from where they might have eventually migrated to the islands and then undergone drift within the species as the result of isolation from the source population—or even as the distinct creation of species on those particular islands. Unfortunately, Darwin was not able to perceive the other possibilities. At this point, one cannot help but ask the following two questions: Is it really possible, or conceivable, that a small volume of sharp rocks could actually create something, or transform life forms into something else? Conversely, is it not possible, and even easy, for a Creator with Infinite Power over everything to create whatever He wills?

It should be mentioned that none of the animals belonging to those islands is as well known as the small land birds which are now commonly known as "Darwin's finches." As mentioned above, the Galapagos Islands are composed of thirteen major islands, six smaller ones, and many islets consisting only of small rocks. In total, there are fourteen distinct types of finches living within this group of islands. Displaying clear distinctions from each other, these birds were classified as fourteen different species belonging to six genera.

The biggest one is about as big as a crow, and the smallest one is about the size of a sparrow. Their plumage is different in color, ranging from light brown to black. Further, the shape of the beak changes from one species to the other; while some have a small conical beak (the *Geospiza* genus), some have a beak similar to a parrot's beak (the *Camarhynchus* genus), and some other groups are comprised of thin-beaked birds, like cornelian cherry birds (the *Cactospiza* and *Certhidea* genera). This variation in the morphology of beaks reflects fundamental differences in both their eating habits and their general behaviors. Some species, having a big conical beak or a parrot-like beak (land finches) are seed and cactus eaters which spend most of their time hopping on the ground. Those that have long, thin beaks (perching birds) are insect-eaters like serins which spend most of their time on tree branches. The species which has a drilling beak much like a woodpecker, and which climbs upright along the trunk of a tree, uses an important feeding technique: it inserts the needles of a cactus plant into small cracks, or slits, in a tree in order to extract insects. Long, thin-beaked warbler finches, which have sharper and more slender beaks, move very fast in a position in which they half-open their wings; in this way, they swiftly hop around bushes and on the branches of trees while looking for insects. Thus, even though they are very diverse in terms of height, color, beak morphology, behavior and food preferences, the fourteen finch species of the Galapagos Islands are assumed by advocates of evolutionary theory to be very closely related. For this reason, according to the artificial classifications which have been done, the finches are included in the *Fringillidae family* by some taxonomists, while they are considered to be part of the *Emberizidae family* by others (as in the zoological encyclopedia by Bernhard Grzimek). As frequently occurs with other animal groups, another taxonomist could still come up with an entirely different classification in the future, whereby she or he may put all of these species into different families or genera. Such revisions are common in taxonomy and will be continuously spotlighted as a result of the discovery and evaluation of new biological properties. Indeed, by no means may we say anything definite about the families or genera into which the birds that are now included in the *Fringillidae* and *Emberizidae* families, for instance, will be classified in the wake of future crossbreeding experiments or new chromosomal studies—nor may we establish with certainty whether they are going to be considered under a new genera, family, or species name. Consequently, in any context where the notions of species and sub-species are being discussed—and given that all of the systematic categories other than species are admittedly synthetic—it is a grossly premature decision to say that these finches came from a common ancestor. The insistence on such a view is a false judgment which does not sufficiently rely on evidence. On what type of evidence, for instance, relies the rejection of the idea that each species came separately from a parent continent?

In the era when Darwin lived, it might have been seen as reasonable to interpret the evidence such that some of the finches living in those isolated islands were deemed related, and originating from a common parent species, since they did, in fact, display a kind of morphological continuity with respect to the shape of their beaks, their height, and the color of their plumage. On the other hand, in a time like today, where the advances which have been made in molecular biology, genetics, zoology, and the migration of the birds, for example, has altered both our knowledge and fundamental concepts, such a claim can only be proposed as a prejudgment for ideological reasons.

Darwin wrote the following: "Seeing this gradation and diversity of structure in one small, intimately related group of birds, one might really fancy that from an original paucity of birds in this archipelago, one species had been taken and modified for different ends."[21] But how did Darwin know whether there had actually ever been a scarcity of finches? How could he know about how those birds initially came to the Galapagos Islands? What made him insist that they could not have come from the mainland separately? If one finch species could reach there from the continent for the first time, could other bird species not reach it as well? Why could the finches not have been created in, or for, the Galapagos Islands specifically? (Note that the classic problem here is mainly due to the fact that those who do not believe in the Creator deem that it is not possible for Him to create and establish whatever He wills, wherever He prefers.) Couldn't some of those that had reached this place have reproduced subspecies or crossbred descendents? (In fact, in this regard, Darwin's finches were greatly exaggerated by Dr. Jonathan Wells in his book, *Icons of Evolution*, a point which we will investigate below, when we argue specifically against Darwin's assertions.) In addition, couldn't some of the same finch species, which had remained on the mainland, simply have become extinct? (In this regard, we need to remember that hermit ibis birds, for example, were at risk of extinction until very recently). Besides, given that conditions are actually not so different from one part of the Galapagos Islands to the other—even by Darwin's own admission—how could such variation among the finches arise as a result of environmental conditions?

In addition to the remarkable variation between species which is witnessed in the archipelago, according to Darwin, there was another aspect of the natural history of the islands which worked against the doctrine of fixism, or immutability, of species: despite the uniqueness of the fauna of the Galapagos, most of the species there were obviously related to sister species on the nearest continent, the South American mainland, located roughly six hundred miles to the east. Darwin commented on this relationship as follows:

> If this character were owing merely to immigrants from America, there would be little remarkable in it; but we see that a vast majority of all the land animals, and that more than half of the flowering plants, are aboriginal productions. It was most striking to be surrounded by new birds, new reptiles, new insects, and yet by innumerable trifling details of structure, and even by the tones of voice and plumage of the birds, to have the temperate plains of Patagonia, or the hot dry deserts of Northern Chile, vividly brought before my eyes.[22]

In other words, while environmental conditions were really quite similar to continental conditions, most of the living species were unique to the archipelago. In fact, this isolated archipelago bore obvious similarities to South America. For this reason, according to notion of "the fixity of species" which was supported by Darwin's opponents at that time, the fauna of the Galapagos Islands should indeed resemble the fauna of South America—but not, for example, the fauna of the Cape Verde Islands, which are actually far closer in climate, geology and general characteristics. The Cape Verde Islands, located on an archipelago near Senegal, in the Macronesia ecoregion of the North Atlantic Ocean, was a necessary stop for Darwin's ship, the Beagle, so that it could catch the trade winds and reach South America expeditiously, as did other ships. Commenting on the various observations he made during his four-week stay in the region of the Cape Verde Islands, Darwin wrote the following:

> Why, on these small points of land, which within a late geological period must have been covered by the ocean, which are formed of basaltic lava, and therefore differ in geological character from the American continent, and which are placed under a peculiar climate—why were their aboriginal inhabitants, associated, I may add, in different proportions both in kind and number from those on the continent, and therefore acting on each other in a different manner—why were they created on American types of organization? It is probable that the islands of Cape de Verde group resemble, in all their physical conditions, far more closely the Galapagos Islands than these latter physically resemble the coast of America; yet the aboriginal inhabitants of the two groups are totally unlike; those of the Cape de Verde Islands bearing the impress of Africa, as the inhabitants of the Galapagos Archipelago are stamped with that of America?[23]

Indeed, Darwin's question was based on the observation of an important phenomenon: namely, if creation in a geographical region is strictly, and ideally, suited to the climate, physical geography, and geological characteristics of that region, then why do the native populations of the Galapagos Islands and those of the Cape Verde Islands not resemble one another? However, Darwin's thinking became shallow, or limited, at this point.

The answer to the question is simply that this phenomenon, which exhibits the richness of creation, is not restricted to the Galapagos Islands. To all well-traveled naturalists, it is obviously apparent that very similar environments on various continents are often occupied by quite different, and unrelated, species. In general, different, yet mainly related, forms populate *adjacent* geographical regions within any greater continental area. So why can the same types of environments not be populated by the same species? First of all, why should they be? Is this not a case, then, which actually proves that knowledge, willpower and planning are not essential attributes, or abilities, of "nature"—which is assumed to possess some sort of virtual power according to evolutionary thinking? Indeed, these questions are strictly logical and have nothing to do with either the belief in a particular faith, or even belief in the Creator.

Darwin was not the only Victorian naturalist whose belief in the fixity of species was shaken just by a trip—specifically, by witnessing the phenomena of geographical variation in isolated regions. Having such a great influence on Darwin's geological thinking through his book, Lyell, who resisted the idea of organic evolution for many years, felt the impact of Darwin's argument after he had been exposed to the phenomenon of geographical variation on the Canary Islands. Also, in 1858, Alfred Russell Wallace, who subsequently proposed the "theory of evolution by natural selection" with Darwin to the Linnean Society, accepted the idea of evolution after he identified a similar phenomenon in Malaya and in the Indonesian Islands.

Static and Dynamic Species: The Secret of Adaptation

In addition to the biological principles mentioned above, the cases witnessed in nature are occurrences of new subspecies, which represent systematic subgroups belonging only to same species and which increase the diversity within a species. Thus, adaptation is a phenomenon which is observable at the end of a process of competition whereby a species is able to overcome difficulties as a result of tolerating the new physical conditions, using its own particular morphological, physiological and behavioral characteristics. A species might have many variants in its new generations. If some of the offspring which are thus produced within the species' genetic capacity do not have the information to code for specific biological activities which are required, or suitable, for that environment, or to sustain themselves in new conditions which may arise in that environment, then those offspring may not be able adapt to the new circumstances and they will perish as a result. Meanwhile, those offspring which have a genetic capacity which makes them suited to the new environment—that is, those which have the physiological mechanisms required for their vital activities, as well as the correct genetic information to operate their organs so that they may adapt to the environment in which they live— will survive and reproduce to yield more offspring which are also favorable to that environment. Yet the occurrence of variations among even those new generations will be naturally evident from time to time. On the other hand, evolutionary theory claims that those small changes which are initiated within the species would exceed the boundaries of the species eventually and thus result in an entirely different species— with diverse genetic material, and which could not interbreed with the previous generation—after a very long process. Of course, being a completely fanciful assertion, neither field observations nor cytological or genetic studies in the laboratory can confirm it.

For instance, as a result of dealing with insecticides, obvious decreases in the sizes and growth rates of insect populations are seen. However, the frequency of resistant genotypes starts to increase in time. As a result of the resistance of these individual genotypes to harsh environmental conditions, populations will often maintain a steady genetic constitution with respect to many traits. This attribute of populations, the ability to reproduce very well-adapted phenotypes, is called "genetic homeostasis." Mosquitoes which have gained resistance against DDT, and bacteria which have become resistant to antibiotics, are very good examples of adaptation. Thus, even though both DDT and antibiotics were quite powerful weapons when they were synthesized for the first time, they have lost much of their previous strength as a result of the high capability for adaptation (coded in the genetic programs) of insects and bacteria. Meanwhile, the resistance of those insects and bacteria that have survived has increased. However, neither the legs nor wings of the mosquito have ever changed, nor has the bacteria transformed into another living being.

In effect, the most important thing at which Darwin wondered during the process of putting together evolutionary theory was the tremendous variation in flora and fauna species. For in addition to a feeling of amazement, it gave him the enthusiasm to search for the source of such variation.

Evidently, the most closely witnessed diversity is in domestic animals and plants—and it is truly striking. A good number of cat races, such as the Angora, Manx, and Siamese, for example, can be counted within the cat species. Similarly, tens of plum and grape species could be mentioned. As a result of such evidence, Darwin came up with the idea of transformation of species such that a great many small differences within one species would eventually accumulate to result in a completely new species. In effect, the change of a grape to a plum, or that of a cat to a tiger—or vice versa—could now be claimed. While none of the proponents of the idea have ever been able to accomplish such a thing, Darwin still believed in the possibility of its occurrence. For their part, neither cultivators nor breeders shared Darwin's optimism because their own experiences disposed them towards the reality that there are restrictions on growing, or steering, diverse animal and plant species. It was simply not possible to break the boundaries which determine the species' characteristics and true nature, though individuals having some differences with respect to partial properties could be bred or grown. If a certain horse type, for example, were bred for many generations—be it small or big, heavy or slight, short- or long-tailed, curly- or smooth-haired, and so on—certain new types of horses could result. But in all cases what would be obtained would still be a horse—not a rhinoceros. Indeed, Darwin, who recognized this problem, alleged that in order for the macro-change to happen, still more micro-changes would have to accumulate over time, so there simply had not passed sufficient time for the macro-changes to occur.

As it turns out, the advances made in production techniques for the past five decades have not given any credit or justification to Darwin's predictions; on the contrary, they have thoroughly shaken what he proposed. Furthermore, the developments in the field of fossil records have brought about additional counter-evidence against evolutionary theory.

The claim that mutations within species somehow become a collective and synergetic power (supporting each other) over time to cause a morphological change resulting in a new species lies at the core of neo-Darwinian synthesis. In other words, this assumption, which necessitates a transition from micro-changes to macro-changes, constitutes the basis of the idea of evolution. However, scientific realities do not support that assumption. Those who conduct improvement, or breeding, studies accept that some changes might occur "within a species," through selectively raising cross-bred animals and plants, and as a result of choosing high-quality strains. Yet starting from the very first pigeon that was studied for selective breeding, all the pigeons which resulted, for generations, were still pigeons—never eagles or even a different subspecies.

Probable or possible supposed improvements have restrictions, then, and they are dependent on laws pertaining to genetic mechanisms. Further, the net effect of these laws governing transformations from the original types is such that the improved species return to their initial forms after a while unless there are deliberate interventions from outside. In other words, selectively produced strains, like huge plants and dwarf animals, have a natural tendency to go back to their original sizes or structures in subsequent generations.

In brief, the emergence of variations within species through the mechanisms of adaptation and natural selection, and which are the result of principles which have been placed in the book of nature by the Creator, only cause a kind of horizontal diversity that we refer to as races or subspecies within the same species—but the idea of vertical change, meaning a transition from species to species, is actually not even a question.

5

Evidence for the Theory of Evolution,
or Preconceived Opinions?

EVIDENCE FOR THE THEORY OF EVOLUTION, OR PRECONCEIVED OPINIONS?

WHAT DO PALEONTOLOGY AND GEOLOGY SAY?

Evolutionists who run after evidence from diverse fields of science in order to promote the evolutionary hypothesis to the level of a theory or law have demonstrated an unbelievable level of skill in distorting each new discovery toward their own worldviews. In fact, should you look at all the fields of science from a certain worldview, and accept that view as the foundation of science in its entirety, you would be able to use all kinds of information by twisting it for the sole purpose of supporting such an idea. And that is exactly what evolutionists have done. Evolution is taken for granted from the start, and all interpretations are forced to strengthen this idea. As new discoveries actually disprove what evolutionists claim, the advocates of evolutionary thought simply and immediately back down from their previous claims, and then buckle down to distort the new information in the same direction, searching for new routes to arrive at their evolutionary ideas. Despite all their efforts, however, not a single serious experiment or observation which could verify the evolutionary hypothesis has ever been presented. The existing assertions have been highlighted over and over again, but all of them have already been disproved. Piece by piece, we can investigate the deficient and misleading information, and the incomplete or corrupted logic, that they propose as evidence according to the scenarios they have been trying to establish in all fields of science from molecular biology to genetics, from anatomy to physiology and embryology, and from general geology to astronomy and paleontology.

"Paleontology" is a field of science which studies the fossil evidence of geological periods. As a discipline of science, paleontology became prominent with one particular concept, the extinction of living forms, and Cuvier was the pioneer of this field. Paleontology began with Cuvier's discovery of some mammalian fossils near Paris which belonged to living forms whose representatives no longer existed.

Cuvier thought that vertebrate fossils indicated the discontinuities of the past, meaning that there were "gaps" between species. In contrast, Lamarck thought that there was continuity throughout fossil history. Cuvier believed that periodic disasters or catastrophes had befallen the Earth; each one had wiped out a number of species, and eventually, such an event had wiped out all life on the Earth. This approach of his would later be called the theory of "catastrophism." Opposing this notion, there were other ideas which relied on the steady accumulation of natural events over enormously long spans of time. They asserted that the geological processes now in operation, and thus directly observable, were sufficient to explain the geological or paleontological remains from the distant past. This concept is what is referred to by the phrase, "the present is the key to the past," and the famed geologist, Charles Lyell, led the movement which was based on this particular understanding. Counted as one of the founders of geology, Charles Lyell was also an advocate of the doctrine of "uniformitarianism," which was initially popularized by him in the eighteenth century, but which was later left behind in second place. Human beings surrender quickly, and they believe, it seems, only in the face of concrete objects that they can see with their eyes and hold with their hands. Being aware of this fact, evolutionists have advanced all of their claims by somehow managing to "give shape to flesh and bones"; as a result, they have succeeded in making their ideas popular and accessible. Distorting paleontological remains by adorning them with imaginary exaggerations, making scenarios about the findings as if they were explaining a truthfully witnessed process, and, at least as important as the first two, expertly employing the mass media for their interests—all these lie behind their success.

A discussion among paleontologists and paleoanthropologists, which would actually require specialized knowledge to fully appreciate or understand, is presented to the general public as if "an important problem related to evolution has been resolved," or "one of the lost links between humans and apes has been found." However, the truth is that what is presented is nothing but an opinion based on a scenario accepted as fact, or a mere debate related to some recently found fossil pieces.

Dating according to scenario

The evolutionary hypothesis also betrays serious problems and contradictions regarding the dating of the age of the Earth and the dating of fossils belonging to various geological times. As we will mention in detail below, dating methods other than those which purport to prove the various ages of animal phyla in a manner which is favorable to the evolutionary scenario are excluded from the literature by advocates of evolutionary theory. For instance, "Rock paintings found in the South African bush in 1991 were analyzed by Oxford University's radiocarbon accelerator unit, which dated them as being around 1,200 years old. This finding was significant because it meant the paintings would have been the first bushman paintings found in open country. However, publicity of the find attracted the attention of Mrs. Joan Ahrens, a Cape Town resident, who recognized the paintings as being produced by her at art classes and later stolen from her garden by vandals."[24] The significance of an incident like this is that it reveals that mistakes can be unveiled only in those rare cases where chance grants us some external method for verifying the dating technique. But what happens in those cases where there is not a firm reference present? In fact, age datings are done according to arbitrary scenarios. Since there are different dating methods, each one having distinct advantages and disadvantages over the other, one might easily enough choose the one which benefits a particular line of thinking while rejecting the others.

As an example of fraud in this field being motivated by the desire for ideological and corporate profit, the activities of Prof. Reiner Protsch von Zieten, of Frankfurt University, should be mentioned. Protsch von Zieten systematically falsified the dates on numerous human "stone age" fossils found in Europe. He dated the fossils as thousands of years older than their actual age. He was also accused of selling the university's skulls for his own profit and plagiarizing other scientists' work. According to *The Guardian* newspaper's report, he even manufactured fake fossils and introduced the ape fossil found in France as if it had been dug up in Switzerland.[25] A committee at Frankfurt University investigated the case and found that "Professor Protsch von Zieten has bastardized scientific truths for the past thirty years." *Der Spiegel* Magazine reported the fraud as follows: "The frauds of an anthropologist at the carbon-dating laboratory at Frankfurt University since 1973, which has dated the ages of hundreds of fossils, falsified the ages of some important fossil samples on purpose...."[26]

Concern about Protsch von Zeiten's carbon-dating estimates arose following a routine investigation of German prehistoric remains by two other anthropologists. Thomas Terberger, of Greifswald University, and Martin Street, of the Research Center for the Early Stone Age, in Neuwied, wanted to check the authenticity of the fossils using modern techniques. So, they sent the fossil samples that Protsch von Zeiten claimed to be from the stone age from Germany to Oxford University for testing. The results which came back from the carbon-dating department at Oxford University were described as a "disaster" by the two scientists. These are important remains that Oxford scientists simply no longer believe to be prehistoric. The female "Bischof-Speyer" skeleton, which Protsch von Zeiten estimated to be 21,300 years old, was only 3,300 years old. A skull discovered near Paderborn–

Sande, in Germany, which Protsch von Zeiten dated as being 27,400 years old, so it was considered to constitute the oldest human remains ever found in the region, is now believed to belong to a man who died about 250 years ago. In addition, the skull fragments called Hahnhöfersand man were not 36,000 years old, as Protsch von Zeiten claimed, but were rather a mere 7,500 years old.[27] Needless to say, those unfortunate evolutionists who had founded their scenarios on Protsch von Zeiten's data, and claimed that that Neanderthal man and Homo sapiens had mated to produce entire generations together, were shocked. Wrongly dubbed the "earliest German," and falsely presumed to be a vital missing link between humans and Neanderthals, Hahnhöfersand man was forced to step down from his throne, since at the time of his existence, as correctly dated, Homo sapiens was already well established and Neanderthal man was extinct.

Further, with his false claims, Protsch von Zeiten caused other scientists working on the propagation of the human population in Europe to make profound mistakes. Due to his fraud, uncountable baseless interpretations about the spread of Neanderthal man in Europe and prehistoric Germany were included as "scientific facts" in anthropology books. Anthropologist Chris Stringer, of the Natural History Museum, in London, aptly summarizes the issue: "What was considered a major piece of evidence showing that the Neanderthals once lived in northern Europe has fallen by the wayside. We are having to rewrite prehistory."[28]

When this deceit of Protsch von Zeiten's was uncovered, some of the fundamental bases of the field of anthropology collapsed, and evolutionary theory was deeply wounded. Also, the following statement of Thomas Terberger clearly shows how evolutionary "theory" was erected on a crumbling foundation: "Anthropology is going to have to completely revise its picture of modern man between 40,000 and 10,000 years ago."[29]

All geological dating techniques are based on the fundamental principle of calculating the rate of some continuous natural processes. One of the most advanced dating methods today is the vibration rate of quartz crystal, which acts by applying electric potential. The best-known example of this technology is in the quartz crystalline watches that many of us wear. Another technique is the decay rate of radioactive elements from the day they were created until today.

Nonetheless, it is not sufficient to have only the dating processes available in our hands. In order to measure the accurate passage of time, three important conditions have to be fulfilled. First, it is necessary to accept that processes remain stable and unchanged, even through times when we are not making any observations. Second, it is necessary to know the beginning value of the clock; that is, we need the correct answers to questions equivalent to the following: "How much water was present at the time when the water clock started working?" or, "What was the height of the candle before it was lit?" Third, it is necessary to prevent external factors from interfering while the process is in operation—just as our electric clock will stop due to the interruption of power if we carry it while we jog outside; in other words, it is crucial to be certain that the conditions under which nature's processes operated in the past did not experience any discontinuity equivalent to a power outage.

In fact, the determination of all these conditions is a problem in dating calculations that we still face today. Since we do not have a technique to observe the times in question—as these have been left in history—or to verify the accuracy of the measurements, we should be absolutely sure that those three conditions all held at the same time in the past, just as they can be ascertained to do so today. Yet, here is where the main problem and disagreements begin.

For instance, let us consider the amount of salt currently in the oceans, along with measures of its influx from the land, to estimate the age of the Earth (as developed by the Irish geologist, John Joly, in 1898). Assuming that the oceans were made up of fresh water in the beginning, and that salt was deposited as a result of land pieces undergoing erosion under the impact of rain—so that the salt contained in them was carried to the seas, and then dissolved in water—this technique seems promising at first glance. Further, assuming that the corrosion rate of land has remained constant until today—therefore equaling about 540 million tons of salt being deposited yearly—this method appears useful. Joly calculated the average salt concentration in the oceans today (about 32 gram per liter) and then the amount of salt in all the oceans (approximately 50 quadrillion tons). From there, he divided the total salt in the oceans (in grams) by the rate of salt added yearly (as grams per year), and thereby estimated the age of the Earth to be about 100 million years.

However, if the three conditions mentioned earlier are insisted upon, the shortcomings of Joly's technique become immediately apparent. First, we cannot be sure that the rate of dissolved salt entering the oceans each year throughout geological times was constant. There is also reasonable cause to think that climate conditions varied a great deal throughout geological times—and included, at different times, ice ages, severe droughts, and extreme rainfalls; this variability could have had an inestimable impact. Second, there could have been some amount of salt present in the oceans in the beginning; in fact, it is not known with certainty that there was not any salt present, and recent studies actually suggest that salt might have entered ocean basins from fused magma under the Earth's crust. Third, it indeed appears that external factors interfered in a process which might have seemed stable. It is now known that huge amounts of salt are circulated again and again in the atmosphere, and new evidence advances the idea that the salt in the oceans might have become constant by now, having reached a kind of equilibrium. For as soon as the salt carried by rivers deposits itself in the oceans, it is transferred to the air via evaporation, and then simply comes down again on land as precipitation. While large amounts of salt evaporate through biological processes, even greater amounts go into the structures of deep ocean sediments as a result of chemical processes that clearly interrupt the normal functioning of our "clock."

In measuring the age of the Earth, all radiological techniques are also disabled by the same shortcomings to a certain degree. "Radiometric dating" techniques, which are used to reach back 4.5 billion years, consist of methods aiming to determine the age of rocks and earth based on the decay of the radioactive elements they contain, which have a very long half-life and thus stay radioactive for a long time. The radioactive elements which are relevant to such studies are uranium and thorium, which decay to become helium and lead; rubidium, which decays to strontium; and potassium, which decays to argon.

The basic principle is this: radioactive uranium-238, uranium-235, and thorium-232 atoms are created in such a way that they can slowly transform into various lead atoms (uranium-238 into lead-206 and helium gas; uranium-235 into lead-207 and helium gas; and thorium-232 into lead-208 and helium gas) over a very long time periods. Critically, the decay rate of each of these is remarkably constant. Unstable uranium and thorium atoms produce alpha particles periodically, yet which atom will decay, as well as when it will decay, is not known in advance. There are billions of atoms in a single deposit of uranium, and thus statistical calculations are required in order to guess the probability of the decay of any particular atom.

The most important part of the theory is that the type of non-radioactive lead—for example, the radiogenic lead-206 which radioactive uranium-238 eventually decays into—is chemically distinct from normal lead (lead-204), which is present in the rocks but is neither radioactive nor radiogenic. Thus, in order to determine the age of a certain rock, the amounts of radioactive uranium and radiogenic lead in the sample are measured. Since the decay rate is known, it is possible to determine the duration of the decay, and in this way, researchers can date the rock in question.

The half-life of one of the most widely used isotopes, uranium-238, is calculated to be 4.5 billion years. This means that half of the given amount of uranium-238 becomes lead-206 after 4.5

billion years. For example, if measurements showed that half of a rock were made up of uranium-238 and the other half were made up of lead-206, this would be assumed to mean that the rock is 4.5 billion years old. However, recent studies have raised important questions about the reliability of this technique.

If the lead formed by radioactive activities is really the last product of radioactive decay *only*, then the rocks in the Earth's crust can be assumed not to have contained any radioactive "parent lead" when they were initially formed—and that might be a respectable starting point for the measurements. However, a closer look reveals that this assumption is actually not valid, for observations and experiments have determined the presence of a separate process whereby "normal" lead transforms into a form which cannot be distinguished from "radiogenic" lead. This transformation occurs as normal lead captures free neutrons. Those neutrons are atomic particles which have the energy to transform normal lead to radiogenic lead (which is a candidate for acquiring radioactivity). In a radioactive uranium seam, some uranium-238 atoms naturally transform into lead-206 as a result of fission (the division of the uranium atom's nucleus into two); and some uranium-238 atoms divide into two by natural fission—and neutrons are released during the process of fission. All these neutrons simultaneously convert the normal lead around them (lead-204) and radiogenic lead (lead-206) into lead-208, step by step. Yet, even with careful experimentation and measurement, this lead-208 isotope cannot be distinguished from the lead-208 which is a radiogenic product of the alpha decay of thorium-232. Critically, while the lead-208 isotope can clearly be obtained in two different ways, evolutionists claim that all the lead-238 isotope which is detected is a radiogenic product of the decay of thorium-232. Therefore, as there is a lot of "radiogenic" lead present, evolutionists assume that the decay process must have been taking place for a long time, and this bends and twists the measurements of the age of the Earth towards the favored

concept of an "old Earth" which the purported evolutionary scenario requires.

Along with lead, the other product of the process of uranium-238
decay is radioactive helium gas with an atomic weight of 4. The total
amount of helium in the atmosphere is supposed to be an accurate
reflection of the radioactive helium which has formed via this decay
process throughout every period in world history. Obviously, if the
uranium-lead dating technique is to be considered reliable, then the
amount of radiogenic helium in the atmosphere has to provide a value
for the Earth's age which is consistent with that arrived at through
measurements of the amount of radiogenic lead in the Earth's crust.
However, the ages which are calculated are so different that they cannot even be compared. For if the Earth is truly 4.5 billion years old,
then there should be approximately 10 trillion tons of radiogenic
helium-4 present in the atmosphere. But there are only 3.5 billion
tons present—thousands of times less than expected.

Some geologists have tried to explain this massive discrepancy
by assuming that the difference—that is, the missing 99.96% of the
expected helium—somehow escaped to outer space from the Earth's
gravitational field, but there is no evidence for this supposed phenomenon. Further, in order to explain the missing helium gas, and
assuming the Earth is really 4.5 billion years old, then the atmosphere would have to lose helium very rapidly—at the rate of about
10^{16} atoms per cubic centimeter per second. However, rather than
losing helium, the atmosphere continues to gain a good amount of
helium each year, as new studies show. The reason is that Earth is
moving towards what can be termed a "thin sun" atmosphere which
is fundamentally based on hydrogen and helium, due to nuclear processes which are occurring on the Sun, and it is simply acquiring
more helium as part of this process.

If we consider the measured amount of helium-4 in the atmosphere now and apply radioactive dating techniques to this, we will
come to the conclusion that the Earth is only about 175 thousand
years old. However, our reliability criteria will still be invalid due

to the possible entrance of helium-4 from outside, which will effectively prevent accurate rates of measurement.

Consequently, the "clock arbitrator role" attributed to radioactive decay is endangered in either case, as the measured value is not the decay rate but rather the amount of decay products—while the exact origins of such amounts are unknown. For this reason, all radioactive dating methods used for the determination of the Earth's age are quite defective and are unreliable.

Along with the problems just described, analytical methods based on the decay of potassium to argon, or of rubidium to strontium, are also riddled with the defects mentioned above. While all of the geochronometric methods developed to calculate the age of the Earth harbor some uncertainties, only one of those techniques—the one based on the decay of uranium and similar elements—renders the age of the Earth in billions of years. Therefore, only this technique is applauded by evolutionists, while the other methods are simply ignored. This is because evolutionists require such a long geological past to prove Darwinian evolutionary theory, in that evolutionary processes are assumed to give results only over a very long period of time. This publicity campaign has been so successful for Darwinians that almost everyone today, including scientists from other fields, believes that the radioactive dating method is the only notable and flawless method among those in existence because of the constancy of universal decay. Yet, these widely accepted beliefs are not actually supported by evidence.

There are many problematic aspects of the methods based on the decay of potassium to argon, or of rubidium to strontium. Critically, potassium minerals are abundantly found in many rocks. Potassium-40 decays after emitting an electron, transforming into argon-40 gas, which has a half-life of 1.3 billion years.

The advocates of the potassium-argon method argue that the argon gas which is formed by the decay of potassium-40 is held in the crystal structure of the mineral formed—"like a bird in cage"— and deposits over time; thus, the assumption is that the deposited radioactive isotope can then be used as a clock when it is measured upon its release. However, the potassium-argon method is uncertain since the final product used in the analysis, argon-40, is a very commonly found isotope which is ubiquitous in the atmosphere and in the Earth's crust and rocks. Indeed, argon is the twelfth most common element on Earth, and more than 99% of all argon is the isotope, argon-40. In terms of physics and chemistry, it is not possible to say whether an argon-40 sample is constituted by radioactive decay, or whether it was present in the structure of the rocks while they were formed. Besides, since argon is an inactive element that does not enter into reactions with other elements, the argon atoms are always retained in the crystal structures of the minerals—whether or not they are radioactive. So it has been calculated that not even 1% of the argon still present on Earth could have originated from radioactive activities if the Earth were 5 billion years old; therefore, at least some of argon-40 in all potassium minerals should most likely have been directly formed as argon from the start, rather than forming through radioactive decay. Therefore, if we insist that radiogenic argon-40 is "a bird in cage," then we have to admit that this cage holds some other birds, too, which have essentially same plumage and cannot be distinguished from the argon-40.

It is important to note that the irregular and abnormal entrance of argon into potassium minerals is not a mere estimate; rather, this finding is supported by many studies performed on volcanic rocks whose ages were first calculated incorrectly. As a case in point, even modern volcanic lavas, which were formed in recent history, have been calculated by the potassium-argon method as being 3 billion years old!

A similar study of the potassium-argon technique, which was done on Hawaiian basaltic lavas, delivered ages ranging from 160 million years to 3 billion years. Then, in 1969, McDougall, of the Australian National University, calculated the age of lavas in New Zealand as being 465 thousand years; however, using carbon-14 dating, a piece of tree found in the lava was dated to be younger than 1,000 years old. The reason for the massive age discrepancy here is the possible entrance of argon-40 into the environment during its initial formation, along with the legacy of argon-40 arising from the source of the magma.

Now let us imagine that the rocks that the samples were taken from were heated again by subsequent volcanic activity. In fact, it is just as possible as abnormal enrichment (i.e., the entrance or gain of argon-40) that those mineral samples could have been abnormally impoverished. Such disordered and disrupted samples will surely render incorrect aging if we only try to apply a simple clock method.

In short, unfortunately, an independent way of verifying the age of any sample has not yet been found. In the meantime, ages which "seem correct" are immediately allowed, as they "give an impression" which is compatible with evolutionary scenarios—that is, with uniformitarianism—and thus, a portable data base is miraculously constructed.

As for radiogenic strontium (strontium-87), it is formed as a result of the decay of rubidium in rocks. However, in general, rocks contain ten times more normal strontium-87 than radiogenic strontium. Thus, the rubidium-strontium technique also raises suspicion since, just as in the case of the uranium-lead method, the same neutron capture processes are at work—only here, strontium-86 transforms into strontium-87 by capturing one neutron.

The most embarrassing aspect of all these different dating methods is that they do not generally give compatible ages for the same rock samples. In an effort to make the ages compatible, numbers are adjusted until they "seem to be correct." Thus, scientists responsible for dating get around the "unreliability problem" by labeling "suitable" rocks to date, and rejecting "unsuitable" rocks from the analysis—their suitability being prejudged according to evolutionary criteria. This practice explains why the results of many dating methods confirm each other—it is simply that all the rock samples which might deliver different ages are rejected as being "unsuitable for dating."

For his part, Richard Milton believes that there are at least four ways by which scientists working on dating get into trouble and error:[30]

First, there are mistakes that cannot be tested. Since independent evidence is not considered, most of the dated ages are not shown as being faulty. In very rare situations where there is independent evidence present, like the cases of the volcanic lavas in Hawaii and in New Zealand, or the case of the paintings by Joan Ahrens mentioned above, the measured ages are found to be surprisingly wrong. The response of supporters of the radioactive dating method to this is that they simply reject those independent verification studies by describing them as a "perversion," and instead prefer to continue to give the credit to their own findings, which are obviously favorable to an "old Earth" view. But while doing this, they throw away the only means of controlling or checking the reliability of dating methods which is available today. It seems, then, that they are so sure about their ideas and their "theory" that they do not need any scientific verification to be done.

Second, events are only considered to happen in their own "playground." Here, a mistake made on the arc of the mirror of the Hubble Space Telescope can be given as an example. Even though the mirror was manufactured in a laboratory equipped with the most advanced technology in the world, the mistake in the arc of the mirror was not discovered by normal control processes. An error in the order of a millionth of a meter could have been found immediately, but a huge error that no one considered checking—amounting to one centimeter—went completely undetected. This was simply because such a big mistake had never even been imagined, and as the measurement criteria had not been set up to work outside the narrow range of what was considered possible, no one perceived the problem which occurred on a much wider scale.

Similarly, with dating methods, the accepted value of criteria has remained within the limits of the "playground" since Charles Lyell first estimated that the Cretaceous period ended 80 million years before today. His ideological colleagues would simply consider any dating expert who offered to check 20 million, 10 million or even 5 million beyond this "playground" to be "crazy." Most importantly, this scientist might not even be able to obtain any funds for his or her research studies.

Third, another reason for potential mistakes is that "minds become locked." The frequent revision of physical constants, which occurs very often, is not well acknowledged. It must be remembered that the speed of light, the gravitational constant, and Planck's constant all underwent important revisions before they became internationally accepted phenomena. One of the reasons for those revisions is that all scientists can make mistakes, and these should be corrected. However, scientists always seem prefer to correct those mistakes with respect to currently accepted realities and values; thus, they give the measured values a senseless, imposed direction. A name has even been given to such a style of thinking—"intellectual deadlock."

Fourth, there is strong professional pressure on scientists to support the generally regarded opinion—the status quo. Because of this, it will be very difficult, and even pointless, for scientists to perform their studies independently or express their ideas freely. For instance, let us consider a rock sample belonging to the end of the Cretaceous period, a period which is believed to designate a time frame about 65 million years in the past. A scientist dating this sample as being only 10 million years old, or 150 years old, could not possibly consider publishing this result, since it would be assumed to be totally wrong. On the other hand, another scientist dating the same sample as being 65 million years old could publicize his or her results widely and publish easily. Therefore, the published dating numbers are always consistent with predetermined ages, never contradictory to them. Should all of the "unacceptable datings" be taken out of the trash can and put together with the published dating results, we would simply be faced with a scattered plot consisting of random numbers only.

Related to the unfolding of dating errors (in spite of the fact that all kinds of precautions were taken, and careful attention to detail was given), Milton summarizes how the individuals concerned could have been urged to do wrong in the following incident, in which even the world's most reputable isotope-dating laboratory was involved.

Paleontologists discovered many human fossils and tools in Lake Turkana (previously known as Lake Rudolph), in Kenya. There was an ash layer defined by Kay Behrensmeyer of Harvard University as "The KBS Tuff" (the Kay Behrensmeyer Site), which was among the significant findings.

When Richard Leakey first started to examine the initial data in Lake Turkana, in 1967, it became necessary to determine the age of the KBS Tuff. Even though it seemed to be suitable for potassium-argon dating, since it was an organic artifact, it was not in its original form (not young); it was corroded, contaminated, carried away by water, and deposited as sedimentary rock. Thus, it contained unknown materials, including odd particles that yielded anomalous ages. Realizing this, the geologists who conducted the dating study chose only the younger pieces from this sedimentary rock formation.

Still, various attempts to date it yielded a wide range of results, ranging from 0.52 to 220 million years of age. Then, in 1969, F.J. Fitch, of Cambridge, and J.A. Miller, of Birkbeck College, London, determined the age of the KBS Tuff to be "approximately 2.6 million years." Later, significant consequences followed from this assertion. For when Richard Leakey found a human skull under the KBS Tuff, he declared that it had been discovered under sedimentary rock "reliably dated" as being 2.6 million years old.

Later, in 1976, Fitch, Miller and Hooker published their second paper on the subject, re-calculating the age that they had determined in 1969 using a more accurate decay rate; they then concluded that the skull was 2.42 million years old. In their study, they ascribed their results to "a small programme of conventional total fusion K-Ar age determinations on East Rudolf pumice samples undertaken at Berkeley Lab."

Another aspect of this matter is that scientists start determining the ages by first selecting rocks that are considered to have the right age, and abandoning samples that seem to have the incorrect age. There is no doubt that this is done obviously and intelligently. Surely, one should be asking the following questions: How do those scientists, working on dating, know which rock has the right age, and which one has the wrong one? What is the reasoning behind the apparent "urge" for them to accept findings of 2.6 million years, for instance—but reject 0.5 million years, or 17.5 million years, in the interest of "being scientific"?

The answer from advocates of dating to these questions is that any scientist would reject a couple of measurements which yield extreme values and would consider instead the majority of numbers gathered together on a "plateau," or in a straight line, when they are plotted. Yet if the measurement process were flawed in the first place, the invariability of the results cannot be support for their accuracy.

Carbon-14 Dating: A Method with Limited Validity

Following the Second World War, in 1949 an American chemist, Willard Libby, made a discovery that secured him the Nobel Prize in chemistry. His invention was truly a landmark in the study of prehistoric periods, but at the same time it turned out to be a development which shook contemporary knowledge and data in regard to dating, and mainly in regard to the age of the Earth.

Libby's invention, which is known as carbon-14 (or radiocarbon) dating, provided the opportunity to determine the age of organic remains. Thus, in the 1950s, field archeologists gave certain ages to the first prehistoric humans, surprising their professors of the former generation by using this new method. It was discovered through this new technique that Neolithic sites in Russia and Africa were actually only about 50,000 years old. In addition, the city of

Jericho, in Palestine, then thought to have been the first human habitation, was deemed to have been established 11,000 years ago.

Archeologists, paleontologists, and especially paleoanthropologists, then began to apply the carbon-14 dating technique to determine the age of organic materials containing carbon (like bones, teeth, charcoal, and so on) which were thought to be younger than 50,000 years

The principle is simple. When cosmic particles coming from space reach the upper part of the atmosphere, they continuously bombard the well-known, stable carbon-12 atoms, which are rich in carbon dioxide (CO_2). Thus, the carbon-12 atom emits 2 neutrons alternately, and radioactive carbon-14 is formed. Yet, carbon-14, which is distributed in an orderly way, is transferred to plants through CO_2 (photosynthesis) first, and then taken by animals as food and thus incorporated into the food chain. There is no difference between carbon-14 and carbon-12 as far as living is concerned: both are common and ordinary forms of carbon which are found naturally on Earth and can be used by any plant or animal, as required. In fact, a living being takes in both of them continuously, in a defined proportion, until it dies. But upon death, while the amount of carbon-12 remains constant, radioactive carbon-14 continues to decay, so that the ratio of carbon-14 with respect to carbon-12 decreases. The determination of the amount of carbon-14 in a sample taken for dating necessitates calculating the decay rate of one gram of carbon in one minute. Since the half-life of carbon-14 is accepted as being 5,700 years, the analyzed organism's death date is calculated on this basis.

Radiocarbon is relatively hard to find; only a small portion of the total carbon in an animal's or a plant's structure is radiocarbon, but it is very simple to make measurements for dating. As soon as it is formed, radiocarbon starts decaying. When an amount of radiocarbon is formed in the atmosphere, half of this amount decays after 5,700 years and becomes nitrogen gas. Then, half of the remaining amount decays in the next 5,700 years; and this goes on until very small amount of residue, which cannot be measured, remains. For instance, after 5,700 years, a tree contains only half of the radiocarbon (compared to ordinary carbon) which it had while it was alive; and after two half-years (i.e., 11,400 years), it contains only one fourth of that ratio. Only an immeasurable residue remains about five half-lives later—or approximately 30,000 years later. For this reason, the radiocarbon dating technique can only be used for the age determination of remains younger than a natural "ceiling" value of 50,000 years, at most. In other words, samples must be younger than 50,000 years for the technique to provide valid results.

The radiocarbon test works on the remains of creatures which were alive at one time—for example, the bones in a grave which is thousands of years old, or the wooden pillars of Roman times. In order to figure out the age of such organic material, it is necessary to measure the amount of radiocarbon left, and from there, to find out when the creature stopped taking in radiocarbon—that is, when it died.

The value of the method unfolds when we need to learn the age of a papyrus piece, for example, or how old a found skull is. In short, the carbon-14 dating technique is based on knowing the true ratio of carbon-14 to carbon-12 on Earth, and most importantly, knowing with certainty that this ratio remains constant in time. In other words, in order for this test to work reliably, the ratio of radiocarbon to ordinary carbon on Earth has to be constant—unchanged since when the creature lived and died, until the time of testing. Indeed, this ratio has been assumed to be constant since the day the test was first developed, but recent advances in this field show that this assumption is incorrect. If archeologists suddenly discovered the grave of a person and would like to date the bones, but there happened to be more carbon-14 at the time the person was living than at the time of the dating test, then the determined age of the bones would necessarily be wrong and that person would seem to have lived more recently than he or she really did. Conversely, if there were less radiocarbon present while that person was living than at the time of testing, then the person would be deemed to have lived further in the past than he or she really did.

While developing this technique, Libby and his co-workers were right to believe that the amount of carbon-14 in the world could not possibly have varied in the course of human existence on Earth, since their estimated time since creation was much smaller than the accepted value of the world's age, at 4.5 billion years. So, Libby considered the radiocarbon rate to be constant as an "equilibrium value" of the radiocarbon reservoir.

According to Libby, there was a 30,000-year transition period required for carbon-14 to become established after the Earth was created and its atmosphere was first formed. At the end of this period, the amount of carbon-14 formed by the influence of cosmic radiation would be balanced to zero by the amount of carbon-14 decaying. In other words, using Libby's terminology and conceptualization, the radiocarbon storage on Earth would reach an equilibrium at the end of 30,000 years.

Yet according to uniformitarian geology (the assumption that the rates and conditions of natural processes operating in the course of geological times are the same as those that can be observed to be operating in present time), since the world is thousands of times older than the time needed for the reservoir to be filled—30,000 years—radiocarbon must have reached a steady state billions of years ago and remained constant throughout the relatively recent time period when humans were created. In order to test such a crucial part of his theory, Libby performed measurements related to both the production and decay rates of radiocarbon and found a considerable discrepancy. For his findings revealed that radiocarbon was being formed 25% faster than it was decaying and disappearing. Since this result was inexplicable according to any conventional scientific means, he simply credited the startling discrepancy to experimental error.

Then, in the 1960s, Libby's experiments were repeated by chemists working with more sophisticated techniques. Since the radiation amount in question is so small (i.e., a decay of a couple of atoms per second), and since it is necessary to eliminate all the other radiation sources which could affect the results, the experiments required very sensitive instruments and measurements. Critically, the new experiments revealed that the discrepancy which was originally observed by Libby himself was not merely an experimental error but rather, an unequivocal fact. Richard Lingenfelter, who verified the discrepancy, commented as follows: "There is strong indication, despite the large errors, that the present natural production rate exceeds the natural decay rate by as much as 25%... It appears that equilibrium in the production and decay of carbon-14 may not be maintained in detail."[31]

These results were confirmed by the publications of Hans Suess of University of Southern California in *Journal of Geophysical Research*[32] and V.R. Switzer, in *Science*[33] along with some other scientists.

A professor of metallurgy at Utah University, Melvin Cook has reviewed the results of Suess and Lingenfelter and has concluded that the present production rate of carbon-14 is 18.4 atoms per gram per minute, and the decay rate is 13.3 atoms per gram per minute; thus, he figured out a ratio indicating that production exceeds decay by about 38%.[34] Cook describes the meaning of this discovery as follows: "This result has two alternate implications: either the atmosphere is, for one reason or another, in a transient build-up stage as regards carbon-14... or else something is wrong in one or another of the basic postulates of the radiocarbon dating method."

Melvin Cook has taken the matter one step further by considering the latest measured data on radiocarbon production and decay, and working backwards to the point where there would have been zero radiocarbon. In doing so, he tried to ascertain the age of the Earth's atmosphere by using the radiocarbon technique. The conclusion he reached, using Libby's own data, is that the age of the atmosphere is around 10,000 years.

The idea that life on Earth may have a history as short as 10,000 years inevitably seems unreasonable to anyone who was

brought up with the teachings of uniformitarian geology and evolution theory—or to any high school student or university student
who opens a standard geology textbook. But has the radiocarbon
technique been tested against artifacts of known age, and thoroughly proven to be valid? Has it been widely verified in archeol-
ogy, with consistent results? Have any fundamental discrepancies in
the technique been discovered?

In fact, radiocarbon dating was attempted on some objects
whose age was independently known from archeological resources,
and it scored early victories. A wooden boat from an Egyptian
pharaonic tomb, whose age was already independently known to be
3,750 years, was one of the first artifacts tested. The radiocarbon
dating trial delivered a date of between 3,441 and 3,801 years, with
only 51 years of minimum error. (One has to wonder whether the
good result was "found" because of the known age of the artifact.)
However, it was right after this promising start that the radiocarbon technique ran into difficulties. Anomalous dates obtained from
successive assays indicated that some creatures might have inter-
acted with certain parts of the reservoir which were deficient in
carbon-14, and thus, they appeared to be much older than they
really were.

Hole and Heizer summarized the situation which resulted from
those anomalous discoveries in their book, *Introduction to Prehistoric
Archaeology*. According to them for a number of years it was thought
that the possible errors were of relatively minor consequence, but
more recent intensive research into radiocarbon dates, compared
with calendar dates, shows that the natural concentration of car-
bon-14 in the atmosphere has varied sufficiently to affect dates significantly for certain periods. Because scientists have not been able to
predict the amount of variation theoretically, it has been necessary to
find parallel dating methods of absolute accuracy to assess the correlation between carbon-14 dates and the calendar.[35]

Being accepted as the oldest living thing on Earth, the Bristlecone pine, which grows at high altitudes in the mountains of Cali-

fornia and Nevada, was used to assess radiocarbon dating by means of comparison testing with a parallel dating method.

The Bristlecone pine has been proposed for developing the science of dendrochronology (dating past events by tree rings) by Charles Ferguson, from the University of Arizona. Since it lives for a very long time, it is very useful—and its particular sequence of tree rings is said to characterize specific years in the past, allowing a younger tree to be compared to older trees (including dead ones) to extend the tree-ring chronology further back, step by step. Ferguson's cross-dating method is used to correlate one core sample to another by way of those particular signatures provided to him, in order to establish a master chronological scale that spans from as far as 8,200 years ago until today, enabling researchers to check the variations in radiocarbon datings.

In turn, Hans Suess has performed radiocarbon dating on the Bristlecone pine based on samples from the master chronology scale and produced a "deviation table" that enables the inaccuracies of the radiocarbon dating technique to be corrected up to about 10,000 years ago. However, a calibration method has not yet been developed for these scales; that is, there is not any settled criteria, or benchmark that we know of from the past to today. The inventor of the radiocarbon dating method, Willard Libby, did not initially think that large deviations would be possible. That is because Libby and co-workers assumed that cosmic rays remain constant—even though they lacked a single piece of evidence to support this assumption. But now we know that cosmic rays fluctuate and that variations do occur in time.

More recently, a new difficulty has been introduced into the controversy. The fundamental principle upon which dendrochronology is based (i.e., that a tree ring forms each year) has been questioned. R.W. Fairbridge, well known for his dendrochronology studies related to the Holocene epoch, states that as with paleontology, certain pitfalls have been discovered in tree-ring analyses. Sometimes, as in a very severe season, a growth ring may not form.

In certain latitudes, the tree ring's growth correlates with moisture, but in others it may be correlated with temperature. From the climatic viewpoint, these two parameters are often inversely related in different regions.[36] Similarly, if the growth starts in the spring and stops due to unexpected cold weather later, and then starts over again, the growth of two rings in one year is also possible, thus introducing further errors into the tree-ring dating method.

The key question here is how to explain the discrepancy between the formation rate of carbon-14 and its decay rate in the atmosphere. In 2001, Warren Beck, of the University of Arizona, and his colleagues, working on the analysis of a stalagmite which started forming 45 thousand years ago in the Bahamas, discovered that atmospheric carbon-14 levels soared dramatically between 45,000 and 33,000 years ago. They proposed that this might have been due to a burst of galactic cosmic rays from a nearby supernova explosion which dramatically increased the production of cosmogenic isotopes.

In that case, if the carbon-14 concentration changed significantly during this period, then dating the fossils of this period becomes impossible. The director of the Radiocarbon Dating Laboratory of Lyon, Jacques Evin, states that the variation of carbon-14 rate over time in the atmosphere has been known for a long time and that is why the determined ages alter frequently. The biggest carbon-14 change observed three thousand years ago makes it impossible to use this method and the other calibration methods like tree rings, coral growth lines and lake sedimentaries' sedimentation limits. He also mentions that the results of this study do not correlate with the results of the bones belonging to the same time period, and he sums up the problem somewhat cynically. For him when archeologists give a sample to dating experts for radiocarbon datings to be done, they are first asked how many digits of a number they expect.[37] In consideration of all the facts, then, we are left with a strong feeling about the unreliability of carbon-14 dating.

No matter how scientific a subject matter is, it is possible only up to a certain point to obtain concrete information or evidence for it in all fields of science. Then, it is left to one's point of view and one's intention as to how to fill in, or complement, the points where it is not possible to secure sufficient supportive evidence. Ideally, scientists should be objective and express only what they have determined by means of experiment and observation—and if they articulate their opinions, they should differentiate their ideas from the definite information or findings. Unfortunately, that is not the case today. Some scientists conduct their experiments with presuppositions about the results, and they look at their findings from that point of view, too. Further, if the experiments or field work do not deliver the desired results, then they completely distort their findings.

Another aspect that one should always keep in mind as a necessity of the nature of science is that what seems correct one day might be disproved the next day—so matters should not be considered settled once and for all. We have seen the fact that even very well-known, "right" findings can be disproved by sounder and more level-headed evaluations. This should be taken into account especially when trying to describe events which occurred in the geological past and which are impossible to repeat. In essence, science has limits, and it is important for those dealing with science to perceive those limits. Yet, as is seen in the above discussion of the problems with the carbon-14 dating method and with uranium-lead dating, evolutionary theory has been progressively losing the very support that it was struggling to establish as its foundation. Therefore, in addition to the fact that it does not satisfy the necessary conditions for being a scientific theory, its existence as an imposition, a value judgment, and a worldview, has gradually become more obvious.

Mass Extinction – Discontinous Creation

George Cuvier first introduced the idea that biological and geological processes did not always function uniformly in the world's history, and did not always display graduality. Rather, those processes had sometimes become more complex, faster, or gotten out of order during major catastrophes, so that it was sometimes a

puzzle how species had existed and become extinct, thus, challenging the fundamental base of evolutionary thought, which was uniformitarianism. Confirming this, geological and paleontological
studies show that life on Earth has not been uniform, but that the
creation of new species after mass extinctions has been observed
from time to time. Despite the speculation about geological age
determinations, most people admit some of the datings which evolutionists propose, which state that starting about 650 million years
ago, mass extinctions happened 440, 380, 250, 210, 65, and 35
MYA, as well as 10,000 years ago. Except for an incident at the
Cretaceous-Tertiary boundary (65 MYA) and the one at the end of
the Permian Period (250 MYA), the three others at the beginning of
the list were stretched over wider time periods, possibly up to 10
million years in duration. Considered in order of their occurrence
from the distant past to more recent time, some are seen as being
related to a flood, and some as small model of a kind of doomsday—the end of the world and life. All these show that geological
and biological processes on Earth have not always had the same
form; in other words, they have not occurred uniformly, for they
have been significantly interrupted from time to time, and big chaotic formations have taken place over a very short period of time.
In other words, uniformitarian thought, through which Lyell and
Darwin tried to manufacture their notion of gradual evolution, is
found to fail.

While doing research in Wales in 1823, the British geologist,
Adam Sedgewick, determined that fossilized sediments were formed
on unfossilized sediments suddenly, not gradually. He named the
period when these fossilized sediments were deposited the "Cambrian
period", and the sediments which were situated below became
known as marking the "Precambrian period." According to the numbers that modern dating methods produce, all the rocks formed in
that given period are accepted as belonging to the Cambrian, even
though some of the Cambrian sediments found in Wales were first
deposited at the beginning of the period, about 540 MYA, and some

were deposited towards its end, about 490 MYA. We mean here to explain the antecedent-consequent relationship in the creation of living beings, rather than to dissect the mathematical accuracy of the numbers related to debated geological ages.

Sedgewick described the beginning of the Cambrian period as a layer featuring the first trilobite fossils found, and this idea has been widely accepted for a century. Note that trilobites, which are thought to have lived between about 550 and 440 MYA, are considered to be the first arthropods and resemble today's crabs. No matter where these are in the world, the places where trilobite sediments are found on unfossilized sediments are accepted as pointing to the Cambrian base. However, today, this limit is considered to be lower, and today, geologists are getting a very good picture of the special "footprint" which marks the beginning of the Cambrian period.

Sedgewick's discovery of such large, complex fossils, which were created suddenly, was certainly trouble for Charles Darwin. In *The Origin of Species*, Darwin mentioned that the Precambrian period was very long and rich with living beings. If that was indeed the case, then where were the fossils of those creatures? If Darwin was right, in order for the complex structured creatures in the lowest layers of the Cambrian to appear, a very lengthy evolution period, wherein primitive "messenger" creatures would have transformed into more complex and structurally diversified creatures, had to have passed. Yet Darwin was never able to disprove this, the firmest criticism supported by evidence which was ever directed toward his theory. Instead, he complained about the missing fossil records, and he expressed a belief in the presence of a series of missing layers under the first trilobite layers, all over the world. He was quite sure that old Precambrian fossils had to be present somewhere. But while the presence of old Precambrian fossils turns out to be true, these are not found in the far distant past, but rather on the Precambrian layers which are right below the Cambrian layers—and they are both rare and very small. Most importantly, they do not have skeletons. In other words, a sudden transition happens from short, nonskeletal fossils to long, skeletal fossils. The formations that declined during the hundreds of millions of years of the Pre-Cambrian Period, and thus the ones that could, or even should, have the missing links between the big phyla, according to evolutionary theory, in fact contain almost no animal fossils. Yet, if transition forms were ever present, their fossils should have been found in countless Pre-Cambrian rock formations.

Today, the Precambrian-Cambrian boundary is calculated to be 543 MYA, and the oldest trilobite fossils are calculated to be 522 million years old. The 21-million-year period between 543 MYA and 522 MYA does not have any fossils in any place around the world; thus, it is named the "pre-trilobite" period. Therefore, according to its accepted age (though the correctness of this age is still debated), our planet was bare of animal life for its first 3.5 billion years. No clear fossil record belonging to the first 4 billion years has ever been found. Nonetheless, as mentioned above, many kinds of bulky animals were created in the oceans about 550 MYA. Still considered one of the most difficult biological events to explain, this time in geological history is known as the "Cambrian explosion." In fact, the majority of representatives of the large invertebrate phyla, which seem decidedly primitive, also first appear on formations representing a very short interval of the Cambrian Period, about 600 MYA. In a geological instant essentially, arthropods, mollusks and some vertebrates appear as the first animals in the fossil record, and our Earth became a planet replete with invertebrate sea life.

More obvious evidence of the Cambrian explosion is found near the small town Addy, in the state of Washington, in the US. It is observed here that there are no fossils present in the lowest levels of thousands of quartzite layers, which are in order on top of each other—but when moving up the levels, the presence of innumerable fossils is suddenly observed (so much so, in fact, that it could be said that the layers abound with fossils). In Addy, there are also remains of Cretaceous creatures similar to small oysters, called brachiopods, as well as sponges and couple of very small mollusks. However, the most common fossil remains found in the first layers there are trilobites, just like in Wales. At first glance, trilobites look like big insects and crabs—but when they are examined closely, they do not resemble any existing living being. The length of trilobite fossils ranges from the microscopic up to 1 meter. They have a large numbers of thorns and heads that look like helmets, as well as distinctive eyes, feet, gills and various jointed legs. Clearly, then, trilobite fossils are evidence of complex and structurally sophisticated creatures.[38]

But if Darwin's evolutionary theory were right, then the first fossils to appear on Earth had to be more primitive than the trilobites. Nonetheless, in many other places on Earth, the very first fossils which are found on top of unfossilized layers are always trilobites—and it is so in Addy, too. Critically, this means that animals with complex structures were created on Earth without evolutionary forerunners.

It was in 1909 that the American paleontologist, Charles Doolittle Walcott, made one of the most spectacular discoveries of an assemblage of new fossil species, which he recovered from the Burgess Shale formation, in British Columbia, in Canada. He found a remarkable collection of wonderfully preserved animals dating back to Cambrian times, about 600 MYA. Along with many well-known animals, such as jellyfish, starfish, trilobites and early mollusks, which were present in these ancient sediments, Walcott found many species that were obviously the representatives of hitherto unknown phyla.[39]

One of the most important of these species was *Hallucigenia*. It apparently propelled itself across the sea floor by means of seven pairs of sharply pointed, stilt-like legs. It had a row of seven tentacles along its back, and each of those ended in strengthened pincers. Another unique form was *Opabinia* with five eyes across its head and a curious grasping organ extending forward from its head and ending in a single bifurcated tip, which it probably used for catching its prey. Being a member of the Chordate phylum, the *Pikaia* was also included among Cambrian fauna which were found in the Burgess shale.[40]

In view of all this information we can conclude that geological studies do not demonstrate an evolutionary change occurring in the layers of the Earth. Rather, a multitude of species of fauna and flora arise suddenly in geological layers and preserve their original structures for millions of years until they become extinct.

Only a very small proportion of all fossil formations were investigated back in Darwin's time, and the number of working paleontologists then was not more than the number of fingers on two hands. Many regions of Earth had never even been walked on, and geologists and paleontologists had examined only a minute segment of the world. Endless regions of Asia, Australia and Africa remained untouched and unexamined by them. Thus, rather than admitting defeat, Darwin insistently argued that an insufficient number of fossils had been looked at; he tried to stand up to his opponents, who rightly claimed that the absence of transitional forms could not be explained by evolutionary theory, and he stated that many of the missing transitional fossils were simply hidden underground, waiting to be discovered. Indeed, he said that it was still possible to find living "missing links" on the undiscovered parts of Earth, but his hope was bound only to the fossils. Thus, the search for missing links continues with fossil formations. Paleontological activities have come to such a point that most studies in this discipline can be said to have been completed since 1860. So, only a small portion of hundreds of thousands of fossil species that are classified today were known by Darwin. Yet all of the fossils discovered since then belong neither to "transition species" nor to "ancestors" of those fossils. Rather, they either look like a species still living today, or they belong to a species which does not resemble any living today, instead representing a completely different systematically categorized species which is now extinct.

Many possible causes for mass extinctions could be mentioned, originating either on the Earth itself, or outside the world. The majority of such major events, namely those at the end of the Precambrian, Ordovician, Permian, Triassic and Cretaceous periods, are thought to revolve around massive fires following asteroid collisions and/or large-scale, periodic volcanic activity—either of which resulted in massive chemical changes in the atmosphere and water, rapid cooling of the air, the halt of photosynthesis, major breakdowns in the food chain, significant temperature and level changes in ocean water on a global scale, and weakening of the Earth's magnetic field as a result of some inversion of the magnetic poles and particular types of weather changes which are thought to impact on seismic activity. It has been determined that collective extinctions especially affected tropical sea animals, and many extinction events overlapped with climatic cooling cycles.

With respect to such calamities, the relative importance of activities taking place outside of the Earth (such as possible periodic phenomena resulting from the particular rotation of the Earth's solar system within the galaxy, including consequences from the activities of the Sun and other cosmic effects) has not completely been defined yet. As witnessed by the fact that 97% of the Earth's rocks are younger than 2 billion years old, what amounts to the continuous renewal and regeneration of the Earth's crust has caused the footprints of geological history to be wiped out. The discovery of only a small volume of fossils of living species, as well as the quantitative and qualitative insufficiency of these in providing accurate data, make it more difficult to understand certain geological events, especially the causative factors behind collective extinctions.

The oldest mass extinction which is acknowledged in the paleontological archives—if, indeed, present age determinations are considered to be true—occurred about 650 MYA, during the Vendian period of Precambrian times. Significant numbers of stramatolites, acritarchs (phytoplanktons) and the soft, multicellular fauna of the Ediacaran period (which takes its name from the region of Australia where it was first designated) became extinct in this point in geological history.[41] Even though this extinction event is not still known well due to the temporal distance clouding age datings and correlations, the possible influence of glaciation has been proposed as a causal factor.

The first crisis which was larger than the one which took place in the Vendian occurred at the end of the Ordovician period, about 440 MYA. Up to 12% of the organisms living in the seas became extinct,[42] and up to 22% of all living organisms are thought to have become extinct.[43] This crisis was connected with a very significant cycle of glaciation which caused massive atmospheric cooling and a considerable drop in sea level. The groups which were most affected were trilobites, graptolites and the first echinoderms—while conodonts, ostracods, chitinozoans, acritarchs and corals suffered only partially.

In turn, another mass extinction occurred at the end of the Devonian period, which ended 380 MYA; more precisely, it occurred at a time in the late Devonian which is termed the "Frasnian-Famennian boundary," 367 MYA. The ecosystems of the seas, particularly coral reefs in tropical regions, were significantly affected by mass extinctions in this boundary. In fact, 90% of all phytoplanktons, all chitinozoans, a significant portion of all fishes, and 65% of all the placoderm species in the seas became extinct. The species living in shallow water were affected more than those living in deep water, and organisms living in tropical regions were impacted more than those living at higher latitudes. In all, 14% of the animal families belonging to the seas were extinguished as a result of this crisis. Important changes in the chemistry of the oceans are proposed as the reason for the crisis; and even though the idea is still lacking a convincing explanation, it has also been surmised that the crisis may have been a consequence of underwater volcanism.[44]

The next mass extinction, at the end of Permian Period, about 250 MYA, is viewed as the greatest of mass extinctions—the most significant and most pervasive of all. Approximately 90% of all the species in the oceans, and more than two thirds of reptilian and amphibian families, suffered extinction in the last couple of million years of this period. Furthermore, the only extinction that insects ever suffered throughout all of geological history happened at this point, as 30% of insect orders vanished.[45]

Recent discoveries of important borderline layers in Italy, Austria and Southern China have shown that the time span of this extinction cycle was much shorter than it was first assumed to have been—specifically, the sudden change which caused disastrous environmental conditions happened much faster than originally thought, and the last crisis phase took less than a million years. It is also suggested that the Permian oceans could have witnessed a very complex extinction model within a very short period of time geologically, according to the Earth's scale, as 49% of all families and 72% of all genera are thought to have become extinct at this point.

The extinctions in ocean environments happened particularly in tropical regions, and reef ecosystems especially were destroyed. Carbon isotopes in sediments indicate a significant drop in the organic productivity of the oceans during this period. Eventually, the oceans became poor in terms of organisms.

This colossal biological crisis at the end of the Permian got the attention of many paleontologists, and various explanations were advanced, from asteroid collisions to global weather cooling. Basically, the main phenomenon associated with the mass extinctions of this period was the significant drop in sea level. According to Anthony Hallam of Birmingham University, sea levels dropped about 200 meters toward the end of the Permian, and continental shelves became exposed. However, this water level decrease was not due to glaciation; rather, it was caused by the continental shelves becoming one single piece ("Pangaea").(36, 37) This might have been the case because Pangaea withheld some quantity of water as an interior sea, and/or because of an increase in the volume of ocean basins resulting from the ridge openings of the middle ocean controlling continent movements.

The mass extinctions at the end of the Triassic period were determined to have occurred 210 MYA in the seas. Most of the ammonites became extinct, and conodonts totally disappeared. While gastropods (a class of mollusks typically having a one-piece coiled shell and flattened muscular foot, with a head bearing stalked eyes), bivalves (mussels), sponges and many sea reptiles were wiped out, significant new creatures were observed, especially among land reptiles. Very important groups incurred massive losses or were completely destroyed during the late Triassic, and their places were mostly assumed by certain groups (dinosaurs, crocodiles, frogs, lizards, mammals, and so on), which appeared in the Jurassic and subsequent periods. Relating the possible reasons for the crisis at the end of Triassic, researchers have advanced many hypotheses, including sea level drops and weather changes. Yet as paleontologist Michael Denton very appropriately points out, the "event" which took place was not only the cause of the mass extinctions but also its result. For, in fact, the emergence of various new groups occurred in living environments which had become vacant by virtue of the extinction of previous forms. The matter to which we need to pay attention to here is that older forms were not eliminated by the pervasive arrival of new ones; rather, those which had completed their duties were discharged by the Divine Power, and new creatures were created especially by Him, with new roles in the life scene.

Biologically, not only all the dinosaurs but also many other groups of organisms played important roles in Mesozoic ecosystems were destroyed at the end of the mass extinctions of the Cretaceous-Tertiary boundary (65 MYA), including two significant cephalopod groups, the ammonites and belemnites; big sea sponges; plesiosaurs and mosasaurs; and flying reptiles, such as pterosaurs, which had remained alive since the Triassic. Other groups were also affected to a certain degree without becoming completely extinct, while a significant reduction in the variety of plankton in the seas occurred.

However, what is particularly interesting is that not all of the groups suffered from the crisis with the same degree of severity; fundamentally, a selective Willpower protected some of the living groups. While land vertebrates like dinosaurs became extinct, most reptiles, such as crocodiles, frogs, lizards and snakes, were not affected much and survived. In general, fresh water animal groups were not overly affected. With regard to mammals, marsupials were affected severely, but placental mammals emerged from the crisis with a relatively light impact. Meanwhile, in the oceans, benthic forms (which live near or on the ocean floor) were influenced less than plankton (which live near the surface of the water); and while ammonites were extinguished, nautiluses survived.

Very significant geochemical anomalies arose in the sedimentary layers of the Cretaceous-Tertiary boundary. Some of these have become the "interpreters" of certain biological phenomena. Two major hypotheses—that of an asteroid hit and widespread volcanic activity—have been proposed to explain the mass extinctions at the Cretaceous-Tertiary boundary.

According to the first hypothesis, known as the "asteroid hypothesis," an asteroid which was 10–15 km in diameter, and which is estimated to have entered the atmosphere at a velocity of 30 km per second, could have hit the Earth and generated an explosion ten times bigger than that which would result from the detonation of the total number of nuclear bombs currently in existence. It has been estimated that the temperature caused by the resulting fire ball would have reached 18,000 °C and destroyed all living organisms in its vicinity as a result of precipitating widespread forest fires. It has been predicted that by effectively covering the face of the Earth in smoke and dust, the cloud rising from the ground as a result of the asteroid hit would have prevented sunlight from reaching the Earth for a couple of months; thus, darkness and cold temperatures (averaging about -30 °C) might have caused the death of plant colonies by preventing sufficient photosynthesis, thereby entailing the death of herbivores.

Conversely, the hypothesis of volcanism rests on the finding of a particular clay mineral, smectite, on a layer where volcanic ash had been deposited over a time span of tens of thousands of years. According to Vincent Courtillot, as witnessed in the last 200 million years of the Earth's history, volcanic activity which drives basalt overland on a massive scale has been observed.[46] The compatibility of substances found in the fresh lava of the Hawaiian Kilauea Volcano, and the amounts of some elements—such as iridium, antimony and arsenic—which are present in the sedimentary layer associated with the Cretaceous-Tertiary boundary, supports the hypothesis of volcanism.[47] During this massive volcanic activity, which is assumed to have lasted for more than a hundred thousand years, poisonous gas spread continually in the atmosphere.[48] This wide volcanic zone, which coated thousands of kilometers in central India and reached up to 2,400 meters in thickness in patches, constitutes the most dense basaltic lava layer in the world.

The mass extinction mechanism of the volcanism hypothesis shows some differences from that of the asteroid hypothesis. First, the extinction event is expanded to a broader time span. Second, the darkening of the sky and cooling are considered to be dependent on large quantities of gas and ash that the volcanoes pulverized—as opposed to the dust and fire clouds which are considered to have resulted from the asteroid hit. Third, the resultant "acid rain" is thought to have been caused by excess of volcanic sulphur rather than atmospheric reactions related to the impact temperature; so, the acid produced would have been nitric acid according to the asteroid hypothesis and sulphuric acid according to the volcanism hypothesis. Fourth, the increase in poisonous gas, causing the death of countless animal colonies through respiratory afflictions, is assumed to be dependent on volcanism instead of huge, sudden fires. Last, it has been proposed that metals like cadmium and mercury mixed into the seas, according to the volcanic hypothesis, thereby poisoning many sea creatures.

However, based on statistical analyses, mass extinctions are defined as cyclic, and they are estimated to have happened once every 26 million years for the past 250 million years.[49] This is interpreted by advocates of the asteroid hypothesis as signifying that celestial bodies periodically hit the Earth and cause mass extinctions. Various hypotheses connecting the extinctions of the Cretaceous-Tertiary boundary and those of other periods with the cooling of the weather follow from these processes.[50]

In addition, some of the advocates of the asteroid hypothesis have attempted to speculate that a shower of comets has hit the Earth, one after the other (rather than just one hit), thus causing extinctions to be spread over time—but this proposition has generally not been accepted.

Furthermore, and considered to be evidence for the asteroid hit, a long-sought crater was found, after many years of searching, in the Yucatan region of Mexico, in 1991; its diameter of 180 km was very close to the predicted size of 150 km in diameter.

The mass extinctions at the Eocene-Oligocene boundary (35 MYA) are named the "Great Break" (or "Grande Coupure"). While some extinctions occurred in the seas during this transitional time, the most significant extinctions were observed among land mammals.

The latest mass extinction event is thought to have happened 10,000 years ago, at the end of the last ice age (the Pleistocene epoch). Animals that became extinct during this period include huge slow-moving animals, such as mammoths, mastodons, glyptodonts, and others. This extinction phenomenon is very well presented in North America, where data shows that an excessive increase in hunting overlaps with the arrival of the first human inhabitants. On the other hand, data in regions such as Africa, Asia and Europe, where human beings had been living for a long time, are not as clear in terms of the scale of these extinctions and their time frames. Overall, interpretations have long sought the reasons behind the extinction of these colossal mammals in the weather changes which occurred as the ice age came to the end.[51]

The Validity of Geological Evidence

The geological evidence—namely, fossil records—is continuously highlighted as the sole witness of the process of "transitioning from species to species," and this is claimed to happen very slowly in terms of both geological and astronomical time. To understand whether a living species such as the ape, whose representatives are still alive today, has gone through changes or not during the geological time period is very hard. It is necessary to perform an accurate paleontological study to come up with a definite judgment as to whether it—or any of its limbs or other features (e.g., its arms, legs, fingers, teeth, etc.) have changed. This is done by the analysis of evidence belonging to species which are completely extinct, like dinosaurs and other species. That is because it is not common to come across fossils that have been totally preserved, and that makes it almost impossible to obtain the necessary information to make comparisons between fossil samples of the same species which lived in different time periods. In order to be able to complete such paleontological research, it would be necessary to undertake the following phases of study:

1) Collecting systematic rock samples, starting from what is older to younger, from fossilized rock formations throughout geological times;

2) Establishing whether or not the fossils belonging to a certain species are common in those rock formations;

3) Ascertaining the number and specific characteristics of such fossils, in the event that they are found; and

4) Establishing whether or not there is a reasonable and sufficient number of fossil samples in each layer to represent the growth stages of the individuals which are among the species to be examined, starting from birth to maturity, for each sample gathered (i.e., a "family picture" should emerge, with infants, youth and elders, all together).

Furthermore, the growth pattern of the individuals of that species from birth to adolescence should be observed in such a family picture. Since this picture would show a family of individuals which represent the growth stages at different ages, starting from birth, it could be defined as a "horizontal cross-section of time." In addition, the changes that the species to which this family belongs have gone through from its creation to today (i.e., throughout its geological time period) should also be mentioned. Evolutionists prejudge the nature of this process by titling it "phylogenetic evolution"—without ever completely understanding whether or not such a living being has actually been through any changes in the past.

In paleontological research consistent with "scientific" methods, the "growth series"—or the populations of the same species on each layer from which the samples are gathered—would have to be determined first. Then, from bottom to top, the comparison between parallel forms—such as baby to baby, child to child, adolescent to adolescent, and old to old—among the fossils representing the pertinent geological ages, for example, from 15 MYA to now, would have to be performed. Only in this way would it truly be possible to claim "scientifically" anything about whether or not any species has been through changes during its geological time scale. In fact, such research and analysis has not been carried out in most parts of the world; and even though there have been some places where there was an opportunity to apply these research methods, it has been impossible to reach reliable results. All these indicate that paleontological—and especially, paleoanthropological—research is not sufficient to explain all of the stages of the history of life; rather, it is inadequate. That is because the fossils found do not provide a chance to perform an ideal study, like the one mentioned above—and the problem is not only related to the quantity of fossils, but also to their quality. Since fossilization is a selective process, the existent fossils are very few in number, insufficient and disorganized. For instance, the fossil samples of invertebrates which do not have any type of bony or cartilaginous skeleton are extremely small in number, and they fall far short of being revealing. Vertebrate fossil samples are also inadequate to explain the changes in species throughout the history of life. Infant and youth fossils cannot easily be preserved since their bone structures are very brittle and so only a very few of them are ever found. As a result, this makes it impossible to understand not only the anatomical differences in species along the axis of horizontal time, but also general changes in the vertical time axis. For example, the total number of human baby fossils found prior to 1998 all around the world is only eight. The last two, which were discovered in South Africa, were of children aged about one and three years old, who lived about 2 MYA (assuming that the dating is correct).

Even more problematic, it is obviously not possible to repeat events that have been experienced in geological history in order to perform experimental observations. Only a minute number of the "footprints" of such events have been preserved reliably so far. Paleontology, or paleobiology, which has played a major role in our search to understand the history of life, has not been able to overcome all of these drawbacks. As a result, objections are simply deemed to obstruct the compatibility of the proposed theories with the criteria of "science." Depending on the data they have, paleontologists establish some scenarios, models and theories, intending to explain the past. Nonetheless, the requirement of "being scientific" is not met—not only in terms of the research methods pursued, but also in terms of the consistency of the theory.

There have been a few studies done to find evolutionary relations between humans and apes among other living species by attempting to apply the above-mentioned method of scientific analysis, but not a single one of those studies has been able to attain a conclusion because the number of completely preserved fossils on which the theory is founded is negligible in comparison to its presumptions. There have been very few human and ape remains, belonging to various ages and different ecological environments, found in different parts of Africa, Asia and Europe. In some cases, there are substantial time gaps, like a million years, between two fossil remains. Further, since the fossils which are found are not completely preserved, and each fossil has many defects, the criteria being used for paleontological analyses and comparisons cannot be standardized. In other words, the fossils cannot be compared in terms of the structure and volume of the skull, projection of forehead, arc of eyebrow, nasal passages, cheekbone, jawbone, teeth, upper and lower arm bones, tibia, thighbone or pelvis. For example, some paleontologists find only a forehead and nasal bone, while some others find a pelvis. Then they draw a conclusion that is far beyond what they would be able to derive scientifically—and thus, they attempt to explain the history of the species. At this stage, where neither distant nor close relevance can actually be established with certainty, ideological preferences take their role. As stated by Geoffrey A. Clark, an expert in prehistoric anthropology and archaeology at Arizona State University, this situation is caused by the fact that scientists who come from different research backgrounds do not share the same paradigms, preconceptions or prejudgments. Thomas Kuhn (1922–1995) comments that each community has traditions particular to different areas constituting its intellectual life. These traditions are located on a base called the "metaphysical paradigm concept." The notion of a paradigm is a way of problem-solving determining scientists' "point of views toward the world" implicitly. The concept of a metaphysical paradigm is, hovever, all of the prejudgment, preconception and preacceptations related to our knowledge of the universe. Therefore, Kuhn argues that it is impossible to come to an agreement in debates, resembling the dialogue of the deaf, concerning the origin of man, and even if new data are found, the problem will not be resolved, because the data are dependent on paradigms, and they are only meaningful within the conceptual framework, which defines them.[52]

In order to show that the discontinuities between the big animal groups can be filled with transitional forms, it is not sufficient to find just one or two types of organisms which have doubtful connections but which then assume the designation of a transitional form in the geological formations being examined. In fact, the correct determination of a fossil organism's status in the taxonomic system, and its biological kinship, is much harder to ascertain than that of a living form, so this can never be achieved with true certainty. First of all, 90% of an organism's biology takes part in the anatomy of its soft parts, and these are not preserved in fossils. For instance, let us assume that all marsupials had become extinct and the entire group was known only through skeletal remains. In that case, who could guess that their reproductive systems are very different from that of placental mammals, and even more complicated than mammals in some aspects? Could we distinguish a marsupial mouse, a marsupial squirrel, or a marsupial wolf from a placental mouse, a placental squirrel, or a placental wolf, just by examining their skeletons? Note that the placenta is a vascular, fleshy, spongy tissue which holds to the uterus very tightly via many points of attachment and connects the fetus with the parent. Except for *marsupials* and *monotremes*, all mammals are placental. *Marsupials* are mammals for which the embryological development in the mother's uterus is fairly short, so the females have an external pouch containing the teats where the young are fed and carried about once the main development after birth is completed. *Monotremes* are a subclass of land and water mammals, having a cloacae (posterior opening) in which the ducts of the urinary, genital, and alimentary systems terminate, and they reproduce by laying eggs. But could we tell anything about the branching of the aorta of an animal that had already become extinct, and for which not a single living individual of that class still survived? Could we learn anything about the unique structure of its heart or kidneys, the shape of its stomach, or the length of its intestinal tract, just by looking at the remains of its skeletal system?

It is worth going further into detail with a simple examination of the contrast between the placental dog family and one particular non-placental predatory *marsupial*. Known as the Tasmanian wolf, and having a dog-like demeanor, the meat-eating *Thylacinus* lived in the open forests and scrublands of the island of Tasmania, very close to Australia, until very recently—only becoming extinct in the 1930s. Even though there is no kinship between this carnivorous non-placental marsupial and the placental dog, they both look so much like each other in terms of their general appearance, skeletal structures, teeth, skulls and other organs, that only an experienced zoologist would ever be able to distinguish them from one another. However, there was a very critical point of divergence between the two groups in the anatomy of their soft tissues, specifically in regard to the placenta—evidence which has effectively vanished through decay and thus did not fossilize. Yet if only their fossils were to be examined, both could be considered to be the same species. The fact that they were actually different species could only be ascertained by comparing the living representatives.

For about a century, the fish belonging to the *Coelacanth* (*Sarcopterygii*) suborder, which are lobe-finned, have generally been thought of as the ideal ancestors of amphibians; therefore, those fish have been classified as being the intermediate forms in the transition between fish and land mammals. This decision was essentially founded on a certain number of characteristics of the skeleton—specifically, the arrangement of the bones of the skull, the position of the teeth and the backbone, and the plan of the fin bones. Also, since *Rhipidistian* fishes actually bear a closer physical resemblance to the first known amphibians, in addition to all the markers mentioned above, it had been assumed that the biology of their soft tissues included a transitional characteristic between typical fish and amphibians.

Yet, in 1938, fisherman pulled a living example of an old *Rhipidistian* ancestor into their fishing nets around the Cape region of South Africa, in the Indian Ocean. The astonishing discovery of this fish, called *Latimeria chalumnae,* of the *Coelacanth* suborder, showed that this species, which was thought to have been extinct for a hundred million years, was actually still living. Since the *Coelacanth* is admitted as a close ancestor of the *Rhipidistia*, the chance to examine first-hand the biology of one of the classical evolution links was obtained.

Finally, the opportunity was available to determine the specific characteristics and functions of a purported ancestor of the vertebrates. This anticipation was based on two prejudgments. The first was that *Rhipidistia* were the closest ancestors of tetrapods; and the second was that Latimeria had evolved from *Rhipidistia*.

On the other hand, the examination of the living *Coelecanth* was disappointing. The largest portion of its anatomy, especially the anatomy of its heart, intestines, and brain, did not fit at all with what was expected of the alleged ancestors of tetrapods. In other words, the modern *Coelacanth* does not display any evidence of having any pre-adapted organs which could be used on land. For this reason, even though the biology of the soft parts of *Rhipidistian* fishes is similar to that of their alleged ancestors, the *Coelacanths*, in terms of their skeletal structures, they actually are very different from the first amphibians with respect to their general physiology. The claim of *Latimeria* evolving from *Rhipidistia* has been seriously criticized by Barbara Stahl, in a broad study of the internal organs, which was briefly alluded to above.[53]

If the case of the *Coelacanth* is evidence for anything at all, it is the reality of how hard it is to reach a conclusion related to the general physiology of organisms just by considering their skeletal remains. Hence, since the biology of the soft tissues of extinct groups cannot be known with any real accuracy, the status of supposed transitional forms—even the ones which may appear to be the most convincing—must be regarded as uncertain.

From the point it has reached today, the study of fossils is challenging to the notion of evolution very strongly. In order to make the big gaps separating the known groups smaller, very many intermediate varieties are needed. In *The Origin of Species*, Darwin emphasized this point over and over again and tried to convince the reader to believe that it is necessary to admit the presence of innumerable transitional forms in advance:

> By the theory of natural selection all living species have been connected with the parent-species of each genus, by differences not greater than we see between the varieties of the same species at the present day; and these parent-species, now generally extinct, have in their turn been similarly connected with more ancient

species; and so on backwards, always converging to the common ancestor of each great class. So that the number of intermediate and transitional links, between all living and extinct species, must have been inconceivably great. But assuredly, if this theory were true, such have lived upon this Earth.[54]

However, talking about continuity based on skeletal fossils causes important problems. In order to support the idea that a big separation in nature does not comprise an impasse of discontinuities, those who believe in evolution as though it were a religion reflect on the similarities in fossil forms with regard to skeletal morphologies with exaggerated interpretations issued for the general public, since they simply cannot talk about soft tissues. Yet, to be able to do that in the first place, continuity should already have been proven by intermediate fossils that would clearly and uncontestedly show the purportedly perfect gradual transition from one species to another. Nonetheless, as Stanley states, the known fossil record is not, and never has been, in accord with gradualism. The fossil record itself provides no documentation of continuity of gradual transitions from one kind of animal or plant to another of quite different form.

According to Pierre Thuillier, the occurring "phenomena" do not give clear and exact answers. The fossils discovered in geological formations do not form perfect, completely continuous series. There have always been gaps and missing links between fossil forms. If one blindly insists on continuity, one could claim that those links only seem to be missing, as Darwin argued. He spoke about the lack of paleontological evidence at that time and claimed that some fossils were simply lost due to some coincidental reasons (or that they had not been discovered yet). But that is not the only possible reason, since both gaps and discontinuities are undeniable realities. In conclusion, the graded evolution scenario, with one species following after another—along with the phylogenetic trees representing the notion of gradualism on which this concept is founded—is seen to be an artificial construct.

Such a viewpoint has been supported not only by Eldredge and Gould, but also by many other scientists. John Sepkoski, from Chicago University, states very clearly that he is tired of listening to people talking about the lack of evidence in fossil records.[55]

The facts described above also extend to plants. The very first representatives of all big groups suddenly appeared on rock formations in complex shapes specially created with many features. Being one of those groups, angiosperms belong to the period from about 130 MYA to 65 MYA, which geologists call the Cretaceous Era. Similar to the sudden emergence of animal groups in Cambrian rocks, the sudden rise of Angiosperms is another case which has resisted explanation since Darwin's time. Angiosperms were created as different groups in such a way that they could survive without undergoing any changes. Soon after their initial appearance, the face of the Earth experienced renewed vegetation within a very short time. Darwin was concerned about this sudden event, and in a letter to Hooker, he confessed that "The rapid development, as far as we can judge, of all the higher plants within recent geological times is an abominable mystery."

Consequently, those examples which demonstrate that fossils can be misleading highlight two evident facts: first, that a major claim such as evolution necessitates strong evidence; and second, that the claim is effectively deprived of such evidence. In turn, the community of biologists is under obvious pressure from evolutionists to assist them in abnegating the existence of God. Other than a few exceptional individuals, the entire community pretends not to see these realities while, unfortunately, the public remains unaware of this desperate situation.

Intermediate Forms

The number of animal species living, given a name, and included in taxonomic systems today is about two million. If the possibility of finding ten million species is accepted, a very simple rationale like the following could be of help in clarifying the requirement that so many species should have left millions of transitional forms behind while "transforming" from a single-celled living form through random mutations and natural selection over time.

For instance, let us think about two species considered to be in somewhat close systematic groups. Let us imagine that there has been a transition between a mole, from the insectivora family of mammals, and a cat, from the predatory carnivora family—or that they came from same ancestor. However, there are almost one hundred differences in the skeletal and muscular systems which can be counted between these two species. Further, should one reflect on all the "smaller" differences in their bodies—like their teeth, digestive tracts, and sense organs—it will become obvious that the number of unique species characteristics reaches into the thousands. Roughly speaking, one could think that "the two species do not differ much," since both animals have two eyes, two ears, four legs, a spine, brain, stomach, intestinal tract, and so on. However, when considered by an animal systematician, namely, when one goes down deeper in the details, the actual differences between the mole and cat will reach up to hundreds of thousands. As a further example, should the feet of a mole and a cat be compared, the special purpose in their structures will be seen in that one is suited to digging the soil, thus functioning as a blade, while the other is suited to hunting prey, thus functioning as a paw. Based on this, the structures and the functions of the bones and muscles display many minute differences. Also, the series of teeth in their mouths are very different: in fact, a mole does not have canine teeth, which are particular to predators. Meanwhile, the sense of sight of a mole, which always lives in a dark environment, does not have the same capacity or operational mechanism as the sight of a cat—even in the same conditions, given the same amount of light. Rather, each species is equipped with distinctive organs and systems so that it may be ideally suited for the environment in which it lives, the manner in which it acquires sustenance, and the specific actions which are necessary to provide for all of its needs. Furthermore, all these differences are present together at the same time, meaning that the individuals of a given species have the opportunity to thrive under the most favorable living conditions, as the present-day situation of every species displays; not a single species which could be called an "intermediate form," and which could be considered to be at a stage of "partial evolution," has ever been witnessed. After all, when we take into account the fact that each diverse organ structure exhibits integrity within the organism to which it belongs, and that each species displays complete unity with, or suitability within, its eco-system, it is clear that such coordination and regulation is a particu-lar preference—that is, a special creation.

Furthermore, if those species really had come from a common ancestor, as evolutionary theory claims, there would have to be dozens of transition fossils, which would necessarily carry many of the characteristics of both species, purportedly showing "gradual" differentiation. The characteristics of those intermediate fossils would differ from each other over time, and cats and moles, which are totally distinct species, would emerge as two separate groups among the most recent fossils. On the other hand, such a scenario has never been encountered in nature. In spite of very deliberate and ambitious studies which have been done continuously for more than fifteen decades, the fossils of so-called "intermediate forms" between cats and moles, or between these animals and their imagi-nary common ancestors, have never been found.

Should the above example be extended to all species in nature, it would logically result in a situation whereby millions of intermediate forms should be available to fill paleontological collections—which are, instead, full of animal fossils belonging to species which are still living today, or to extinct species like dinosaurs. Among those collec-tions, we have never seen a single fossil that displays transitional characteristics. Even though it is easy to draw on paper the figures of a flying mammal, such as a bat; or a running mammal, such as a deer; or a swimming mammal, such as a dolphin; or a climbing mammal, such as a sloth; or a digging mammal, such as a squirrel, and so on—

and to somehow "unify" them under a shared ancestor by indicating a reference back to the past with dashed lines, it is actually not possible to show individual representatives of such drawings or any of the hundreds of transition forms which are presumed to be present between animals with supposedly common ancestors.

Above, as an example, we gave two animals which are included in the same class (mammalian), so the basic functions of most of their systems—such as respiration, circulation, excretion and reproduction—show similarities. But when we imagine the radical differences between certain groups with respect to these vital functions, which are critical for each organism's optimal functioning within its own distinct ecosystem—like fish and frogs, frogs and lizards, or lizards and birds—it becomes clearer how careful one should be when speaking on this matter. On the other hand, keen advocates of the theory of evolution seem to think that it is easy to say that a running lizard, who somehow "understood" that it would not be able to catch insects while running, started to develop wings by essentially "stunting" its front and back feet and its long tail, and "acquiring" a beak from a different material entirely, somehow, and shortening its tongue. Evolutionists make these claims on behalf of science, expecting both their students and the public at large to imagine such nonsensical ideas right along with them.

According to David Raup, curator of the Field Museum in Chicago, where examples of 20 percent of all discovered fossil species are kept, the evidence does not at all support Darwin's contention of gradual, step by step evolution, with numerous intermediate forms bridging one species to the other: "Most people assume that fossils provide a very important part of the general argument made in favor of Darwinian interpretations of the history of life.... Well, we are now about 120 years after Darwin, and knowledge of the fossil record has been greatly expanded....ironically, we have even fewer examples of evolutionary transition than we had in Darwin's time."[56] In fact, the *non*-existence of transitional or ancestral forms in all those fossil crates is accepted as one the most striking properties of fossils by paleontological authorities. In a publication by the British Museum, it is stated that none of the fossils examined is the ancestor of another.

In terms of general characteristics of fossil formations, there are striking gaps between phyla, classes and orders, and new sorts suddenly appear in environmental settings. It is very interesting that fossils in layers of sedimentary rocks arose as quite complex and perfect structures. Medusa (jellyfish), molusca, porifera, arthropods, crustaceans, and many other invertebrates existed together during the Paleozoic Era. So, it would be essential for evidence of evolution for transitional fossils indicating ancestral forms to be widely found in rock formations dated to pre-Paleozoic times, but this has never been observed. Aware of this failure, American paleontologist, G. G. Simpson, confessed his reservations in 1961, upon his examination of the fossil record, as follows: "It remains true, as every paleontologist knows, that most new species, genera, and families and that nearly all new categories above the level of families appear in the record suddenly and are not led up to by known, gradual, completely continuous transitional sequences."[57] This admission obviously demonstrates that there are not any intermediate fossils in evidence in the transition stage. However, at the same time, Simpson discussed fossils in such a way as to deliberately emphasize the occurrence of gradual transition in some aspects in his book, and the answers to many of the following questions will address how those supposed "transition fossils" are indeed invalid and misleading.

From Fish to Amphibians

The origin of fish species, and the subject of their possible ancestors, has also been a continuous mystery for evolutionists who do not want to acknowledge creation. According to the present fossil record, most known fish groups seem to have arisen within a very short time interval about four hundred MYA. Upon their initial appearance, they, too, were separate and isolated with respect to previous living groups. None of the fish groups introduced by paleontology are classified in a way such that one is viewed as an ancestor of another—rather, all of them have the same "value," meaning that each of them is neither an ancestor nor a descendant of another. Thus, the Lord of the worlds, the Lord of all classes of beings, manifests His infinite knowledge, wisdom, will and power by creating innumerable creatures, both in quantity and variety, as the essence of His art.

The absence of transition forms in fossil formations is also clearly proven by another specific group which has unique characteristics that its supposed ancestors did not possess—amphibians. Let us consider the proposed transition from fish to amphibians (organisms that can live both on land and in the water, such as frogs, toads, and salamanders) according to evolutionary theory. The differences in their structures and functions are numerous, and even the occurrence of a small change would have taken millions of years; therefore, surely, innumerable linking forms between fish and amphibians should have emerged in the meantime—as evolutionary theory would necessitate. However, not a single representative of such proposed "bridge forms" has ever been found anywhere on Earth.

We understand from the fossil record that many old amphibian groups, whose representatives have long been extinct, existed for a period of about fifty million years, about three hundred MYA. The first amphibians had the front and back feet of a normal tetrapod, which made it easy for the animal to move over land. Thus, it was ready for life on Earth from the start—in other words, it does not represent a transition to a living form. Once again, each group is isolated and different from the other right from the first emergence, and thus none of the groups can be considered to be an ancestor for another.

Furthermore, there is also a basic difference between the anatomy of all fish and all amphibians which is not linked by transitional forms: the pelvic bones of all species of fish, living or fossil, are small and closely embedded in muscle, and there is no joint between the pelvic bones and the vertebral column. This is because there is no need for the pelvic bones to carry the weight of the body in fish, as water provides the necessary support. On the other hand, in tetrapod amphibians, living or fossil, the pelvic bones are very large and firmly attached to the vertebral column; this is the type of anatomy which an animal must have in order to walk. But there are absolutely no transitional forms of pelvic bones which are evidenced between fish and amphibians.

Instead, the fossil records show that between the fin of crossopterygians (lobe-finned fishes) and the foot of amphibian *Ichthyostega* (the earliest true tetrapods), there is an anatomical gap so large that it makes one ask the most basic question once again: Where are the millions of intermediate forms that would be required to exist in order for the former to evolve into the latter? The links are nowhere to be found. The first amphibian was created in such a way that it could move easily overland with four normal feet—two at the front and two at the back.

From the Land to the Sea / From the Sea to the Land

Animals such as seals, manatees, dugongs and otters, which are either fully aquatic or partially aquatic mammals, are specialized representatives of different groups, and none of them can be the ancestor of today's whales. We would have to force ourselves to assume the existence of many species, totally extinct, in order to reduce the gap. Evolutionists start this series with a small land mammal, an insect-eater about the size of a mouse, and they propose certain "phases" from otters, to seals, to dugongs, respectively, until they eventually reach the imaginary ancestor of modern whales. At this point, it is necessary to imagine many primitive whales to fill the significant gap at the branching area where toothless whales are distinguished from the toothed ones. According to evolutionary theory, departing from being such unspecialized landforms, those imaginary species series must have caused sub-branching. This is because the rationale which lies behind this "theory" is actually "random branching." However, none of the above-mentioned animals is sufficiently primitive to allow for mere coincidence to have steered its development. Therefore, the reality of creation comes to mind once again, for the Creator, Who has power over and knowledge of everything has created all living beings with wisdom according to His particular preferences. However, this idea is totally contrary to the spirit of Darwinian theory, for it ruins any attempt at proposing a firm mechanistic explanation for the history of living beings. Yet Darwin's idea necessitates the existence of innumerable sub-branches causing many unknown species to emerge, and the presence of many more species between the gaps than those that could have emerged if evolution had followed the shortest path. Darwin simply countered that some of those species could have been eliminated by natural selection and the remaining ones would gradually have been "transformed" into sea mammals. Such a dream was so beautiful that he was reluctant to abandon it, but it had nothing to do with reality.

In fact, in order to change a land mammal into a whale, there have to be countless changes in a great number of organs and systems. The following are just some of the major alterations required: modification of the back feet; improvement of the tail fins; appearance of a new profile; shortening of the front feet; transformation of the skull to permit nostrils to come to the top of the head; change in the trachea; modification in behavior; altered functioning of the kidneys to allow survival in salt water; the formation of special nipples to allow newborns to be fed under water; the complete change of the birth process; and so on. To account for all those changes, we would have to consider the existence of thousands of transitional species along the shortest path from the imaginary ancestor, living on land, to the common ancestor of the modern whales.

Life on land has its own particular living conditions, as does life in seawater and life in freshwater. On land, the body faces the danger of losing water and drying out. For this reason, its skin is protected by a hard, dry, keratin layer, which prevents the body from losing water. Also, in order for land animals to act against the force of gravity, they also have to have stronger legs. In the sea, there is no danger of drying out, but animals are exposed to a lot of salt entering their bodies (as in the case of sea fish) or to losing excessive amounts of salt (as in the case of freshwater fish). In addition, the necessary hydrodynamic body and shape of their fins for swimming have to be different from the shape of a leg which is useful on land. Indeed, even we look at them only in terms of their outer morphology, we see that each animal is created with such design and wisdom that each and every property of animals not just allows, but enhances, their belonging to two different mediums—and that from the glands on the skin to the dissimilar muscles of fins and legs, all of the conditions of the medium in which the animal is living are taken into account.

The genetic program coded in the DNA determines even the tissues relating to the smallest organ. First, all of the changes which can possibly occur have to arise in the form of information, either in the animal's zygote (impregnated egg), or in the sperm and egg, separately. For example, for even only the kidney nephrons to change to structures totally different from those suitable for life on land, complete knowledge of the entire structure of an animal is required, and a very strong Power is necessary to put this knowledge into practice. We are currently able to understand how one characteristic transforms into another in accordance with other characteristics, and without changing the whole genetic system, only through the application of modern knowledge of physiology; however, the evolutionists' only basis for the transformation of one characteristic into another is the concept of random mutation. Yet, we are incapable of calculating how many well-directed, controlled, and successful mutations would be required for only the kidneys to change. This is because the occurrence of any random mutation in the kidneys only damages the normal functioning of the kidney and puts the life of the living form in danger—or, at best, it yields no visible improvement since the co-ordinate changes in the other functions or aspects of the kidney would not simultaneously occur.

Nevertheless, a change in only the kidney tubules would not be sufficient either. Requiring numerous critical calculations, it is necessary for other essential changes to also occur for the successful transition from land to water, such as in the structures of the respiratory system—including the lungs, heart vessels and brain, and all other functions and organs responsible for respiration. All of those changes have to happen at the same time since, if all the changes which have an effect on a system and which are supposed to transform it from one level to another do not occur together, that system cannot continue its operation. Therefore, the necessity of the co-occurrence of hundreds of exact and successful mutations on the DNA has to be considered.

If we accept that so many random mutations could occur overnight, it means that we should also accept that a bird might come out of a lizard egg, or a cow might give a birth to a seal. Seeing that such a proposal is very tenuous indeed, evolutionists, willingly or not, are obliged to conclude that transitions must be extremely gradual. On the other hand, in order for the transitional living form to survive at each stage of such a gradual transition process, that living being would have to come to life with the precisely necessary organs—neither with missing nor additional organs. However, in this case, such a living being really could not be called a transitional living form because in order for it to be considered a transitional form, some of its characteristics would have to belong to the previous form, while some of the characteristics would have to be completely original, belonging only to itself as a new form. In such a case, there would emerge a very difficult problem: the compatibility of two distinct models within the same system. Besides, such changes, evidently directed towards a certain goal, would have to be pursued with a totally embracing Willpower and consciousness—yet evolutionists do not admit such a possibility in the least.

Nonetheless, if we try to explain the well-directed changes which are presumed to occur in transitional forms by relying on the notion of successful random mutations, we have to at least acknowledge how uncertain the emergence of information coding only the most suitable structure among millions of possibilities actually is—and how it is virtually impossible to ensure the compatibility of the change in genetic molecules relating to this information with the existing genetic code.

Transition from Invertebrates to Vertebrates

One of the biggest problems is the lack of explanation about the transition from invertebrates to vertebrates, since invertebrate and vertebrate animals are totally different from each other in their body structures and organs. This difference is so big that it is essentially impossible to fill the gap between them with intermediate forms that "improve" gradually. Most invertebrates, like arthropods, echinoderms and some mollusks, have an outer skeleton surrounding the body, like a kind of coat made of chitin or calcium carbonate. Some of them, like annelids and coelenterates, and most of the small phyla are soft animals without a skeleton. Vertebrates have an interior skeleton made of bones or cartilage. Due to the differences in skeletal structures, the muscles are created with such a design that they wrap the outer skeleton from inside in invertebrates, and they wrap the interior skeleton from outside in vertebrates. For this reason, the proposed transformation from an invertebrate to a vertebrate would necessitate an inversion process that would fundamentally turn the animal inside out, so to speak.

Besides, a gradual transition between the central nervous system of vertebrates and the rope-ladder-like or diffuse nervous systems in invertebrates cannot even be imagined. Similarly, there are very many differences requiring big changes in all other major systems. There are indeed so many differences in their organs, of which the following are only a limited set of examples: invertebrates have open circulation, while vertebrates have closed circulation; invertebrates rely on nephridial organs like tubules for excretion, while vertebrates require kidneys; invertebrates have a one-layered body cover, while vertebrates have a two-layered skin;

invertebrates have a trachea and ectodermal gills and use the extended surfaces of their bodies for respiration, while vertebrates have lungs or endodermic gills. All in all, even these few comparisons make it impossible to conceive of transitional fossils between vertebrates and invertebrates which would have their organs "improved" through random mutations. Furthermore, such fossils have never been found in practice.

From Reptiles to Birds

Nowhere else are the shortcomings of evolutionary theory more pronounced than in the case of birds. One of the most prominent experts of the subject, William Elgin Swinton, was forced to accept that, once again, "There is no fossil evidence of the stages through which the remarkable change from reptile to bird was achieved."[58] Nevertheless, after straining their imaginations, paleontologists considered a potential candidate which could be thought of as an intermediate form. The news about the finding of an intermediate form called *Archaeopteryx* (a bird which is said to resemble a reptile) was greeted with joy and cheers. Even though it was definitely a bird, with all of the requisite characteristics—such as wings, feathers and flight—by taking such features into account as its teeth, the vertebrae along the tail, the dense bones, and tiny claw-like appendages running along the edges of the wings—a resemblance to reptiles was ascribed to this species.

First, however, it should clearly be pointed out that the reptilian-like features found in *Archaeopteryx* were more cosmetic than structural. For instance, the presence of teeth in *Archaeopteryx*'s mouth, considered to be a similarity with reptiles, is actually not one of its main features but rather a kind of "detail." This is because there are toothless species among toothed fish, and there are toothless species among amphibians (for example, among some land frogs, like *Bufonidae*), as well as among reptiles, like turtles. Even some mammalian groups, like *Edentata*, are toothless. Therefore, even though modern birds are generally toothless, toothed species could have lived in the past. Thus, being toothed, or not, is not expressly considered to be an essential characteristic of a class of animals; rather, it is the kind of feature which shows differences within the same class. Besides, living birds with claw-like appendages (such as *Opisthocomus hoazin*) have been found since the discovery of *Archaeopteryx,* thus casting doubt over the inflated importance attached to this single creature. All other considerations related to *Archaeopteryx*, which were once thought to be significant, are no longer seen as being important because the case of *Archaeopteryx* was finally put to rest in 1977, when *Science News Magazine* reported the discovery of a new bird fossil in rock formations belonging to the same geological period, demonstrating that the so-called "missing link" lived and flew side by side with other birds, thus precluding the possibility of its being an ancient ancestor.[59] Indeed, *Archaeopteryx*, thought to be 150 million years old, was just another bird—not the most attractive representative of birds, perhaps, but still functionally very much a bird.

Even though many paleontologists have dismissed the claim that it is an intermediate form, *Archaeopteryx* is found gracing biology textbooks with its toothy smile. Another discovery, which further reduces the potential "evolutionary value" of *Archaeopteryx*, is the fossil belonging to a bird dated to 225 MYA, *Protoavis texensis*, which was relatively recently found by Chatterjee, in Texas, in 1991. *Protoavis* represents a flying bird complete with feathers and hollow bones, just like birds living today—and yet it is 75 million years older than *Archaeopteryx*. Therefore, it can be concluded decisively that *Archaeopteryx* can neither be an ancestor nor an intermediate form for birds. In addition, this bird could not have evolved from dinosaurs either, since it was older than dinosaurs. Furthermore, *Archaeopteryx*, which was said to have derived from bipedal carnivorous dinosaurs (theropods) and which was then placed in the "ancestral seat" of birds by means of pragmatic evolutionary rationale, was actually not just different from either species in terms of "details," but also in terms of substance. Even though there were holes in the thigh areas of both groups and in the lower parts of the bones, making the skeleton lighter, *Archaeopteryx* did not have those holes. In addition to that, the respiratory systems of birds and dinosaurs did not have any similarities whatsoever.[60] In China, the discovery of fossils belonging to birds known as *Confuciusornis sanctus*, in 1995, and of *Liaoningornis longidigitris*, in 1996, deadlocked the evolutionists entirely. *Confuciusornis* was toothless, like today's birds, and it was said to have lived 140 MYA, in the Cetacean period. In addition, it was not substantially different from today's birds with respect to the last part of the vertebrae, having a distinctive bone structure called a "pygostyle" and feathers. As explained in *Discovery Magazine* by the famed ornithologist, Alan Feduccia, from North Carolina University, *Liaoningornis* was estimated to be 137–142 million years old. In addition, its breastbone, to which its flight muscles were connected, was similar to that of today's birds— though it had teeth. The importance of the *Liaoningornis* fossil is that it makes a clear case that dinosaurs were not ancestors of birds, as Feduccia argues in detail. Even the bird fossil known as *Eoalulavis*, estimated to be 180 million years old, was older than *Archaeopteryx*— yet its flying was masterly, as could be clearly understood from its body structure.[61, 62]

All of these points aptly demonstrate that *Archaeopteryx* is not an intermediate form; rather, it is a bird species, which lived during the same period as today's birds, along with some other extinct forms with specific structures. Ultimately, then, the common presence of certain characteristics in species belonging to various genera does *not* prove that those species derive from each other. The extinction of a number of birds (toothed ones); the evidence of different structures; and the survival of other birds (toothless ones) until today—all these do not combine indicate that one had come from the other. Rather, they lived together during the same period of time.

As a matter of fact, *Archaeopteryx* was an excellent flyer—which is, after all, the most characteristic feature of birds. To ensure its successful flight mechanism, there are feathers on its wings which are as developed as any modern bird's feathers, and research has shown that these feathers were even capable of performing propulsive flight.

Of course, dinosaurs are not alive today. We could never even have imagined such huge animals, weighing 120 tons, and measuring up to 7 meters between the heart and brain, if we had not found their fossils. And yet, we somehow expect all bird that existed in the past to be exactly the same as the ones living today. However, birds like *Archaeopteryx* did live in the past, and they became extinct, just like the dinosaurs. The Creator does not have to obey the bird model present in our minds in order to create birds. By creating hugely diverse types of birds instead, He manifests the reality that He is the Most Powerful, and that it is easy for Him to create so many varieties.

As the brain is a soft tissue, when fossils are examined, predictions about some of the features of organisms are made only by using the volume and the morphology of the skull. In order to do that, an endocast of the inner cavity of the skull, showing the approximate shape and circumference of the brain, is prepared. According to the endocast of the inner cavity of the skull of *Archaeopteryx*, its brain essentially looked like brain of a bird, with respect to all of the major sections. The brain hemispheres and cerebellum (related to balance and critical movement coordination) were like those which are typically present in the brains of birds. Note that with respect to the size of the entire body, the cerebellum is bigger in birds than in all other vertebrate classes, and it is considered to be a necessary center that plays a critical role in the control of very complex motor movements. In fact, the presence of a big cerebellum in a bird's central nervous system adds new evidence for the hypothesis that *Archaeopteryx* was able to execute active flight, just like today's birds. As a matter of fact, this hypothesis is also confirmed by similarities in its wings and parallels to the firm wing feathers in today's birds. If *Archaeopteryx* had such an ability, then by the same token, would it not also have had the required, nervous, respiratory and circulatory systems, which could supply enough oxygen for the increasing need of active flight? In other words, could it not have been as much of a bird as any other bird with respect to all of its major anatomical and physiological characteristics?

The reptiles and birds living today show major differences in their anatomical and physiological features, especially their nervous and respiratory systems. Since it is not possible to obtain information about the physiology of the soft body parts when starting from the skeletal remains of a fossil form, knowing how much of a bird *Archaeopteryx* actually was with respect to its main physiological systems will never go beyond the level of guesswork.

Some experts have defined the estimated parents of the closest ancestors of three big flying vertebrate classes, namely *Pterosaurs* (now extinct flying reptiles), birds and bats; however, there is a big gap between each of the first representatives of those three flying classes and so-called similar types.

David B. Kitts, Professor of Geology at the University of Oklahoma, summarizes the evidence against evolutionary theory when he observes that evolution requires intermediate forms between species and paleontology does not provide them.[63]

From Reptiles to Mammals

Misleading results will also arise due to rash decisions made just by looking at some reptile fossils, apparent "bridging forms," and declaring them to have skull and chin morphologies close to those of mammals. The possibility that those reptiles, which are claimed to resemble mammals, were actually fully reptiles in terms of their anatomy and physiology can never be dismissed. The only hint about the physiology of their soft body parts which is in our hands is the endocasts of their internal skulls, and those endocasts lead many to think of them as being fully reptile in terms of their central nervous systems. For instance, regarding their purportedly "mammalian" reptile brains Jerison, who is an expert in investigating the endocasts of this type of fossil species, reports that these animals had brains of typical lower vertebrate size; since their endocasts were all very near the volume of these expected brain sizes and since the endocasts present maximum limits on their brain sizes, the mammal-like reptiles, were reptilian and not mammalian.[64] In short, mammal-like reptiles are reptilian and not mammalian in terms of the shape and size of their brains. As a matter of fact, Jerison also fails to say anything convincing about how complex centers, such as those for smell and vision in mammalian brains, could have ever differentiated in such an orderly way by means of random mutations.

If we briefly explain some of the structures required to transition from reptiles to mammals, it will be better understood how impossible it is for such a process to occur. First of all, reptiles, whose bodies are covered with firm, shiny, keratin flakes and scales would have to lose these features through the transformation of those flakes into hair or fur. But surely, this process alone is not sufficient to do the entire job. Essential features of the skin—such as glands for perspiration, fat tissues, and milk glands and ducts— would also have to be developed. Perspiration (sweat) glands are required to help in heat regulation, water stability, and the excretory system of the body. In turn, milk glands and ducts are essential for providing a food supply for the offspring. Can we even fathom the absolute unlikelihood of the profound improvements which would be required to engender such spectacular structures on a reptile—each of them belonging only to the skin—occurring simply by chance?

Furthermore, there is only one bone in the jaw of mammals, and the teeth are placed into the hollows of the bone. Being diverse (heterodontic) in shape and length, mammalian teeth include incisors, canines and molars (including both premolars and molars). However, there are at least three or more different bones present in the lower jaw of each different group of reptiles (turtles, lizards, snakes, and crocodiles). Except for crocodiles, the teeth do not reside in the hollows of the jaw in reptiles; rather, they are just stacked loosely on the jaw. In contrast, turtles do not have any teeth. Except for adders (a type of snake), most toothed reptiles have teeth which are all of the same type (homodontic).

Consider, too, that there are no temporal fossae (cavities) in the cheek region of the skull of some of the different reptilian classes, such like turtles; some of them, like extinct dinosaurs from the *Synapsids* have only one temporal fossa; and the others, such as snakes, crocodiles and lizards, which are placed in the *Diapsids*, have two temporal fossae. On the other hand, the temporal fossae of mammals are wide and large, and support the strong jaw muscles. In addition, the middle ear of all reptiles has only one bone called a "stapes." Contrary to this, there are three tiny bones called ossicles (the malleus, incus, stapes), providing the connection between the ear drum and the inner ear, present in the middle ears of mammals. Keep in mind that it is vital for those ossicles to be joined and connected next to each other without touching, at particular angles, in order for the hearing process to occur in the best way. Is it possible for a reptile on its own, lacking conscious control over its structures, to develop those three bones in such a perfect way? Could "nature" really generate such a well-measured and perfectly arranged mutation by itself?

While the skull is joined to the cervical vertebrae only by one bulge, called the occipital condyle, in reptiles, it is joined to the cervical vertebra by two occipital condyles in mammals. Both the male and female urogenital systems of reptiles are also very different from those of mammals, since reptiles reproduce by laying eggs. A common channel in the male reptile carries both sperm and urine, but sperm and urine channels in mammals are separate from each other. All the necessary conditions for the embryo to develop and grow are prepared in the uteri of mammals. A specialized organ called the "placenta" develops in the uteri of placental mammals during pregnancy. It is connected to the baby via an umbilical cord and supplies all the nutrition the baby in the uterus needs. On the other hand, reptiles reproduce by laying eggs outside their bodies, leaving them to rest somewhere else, in or on the ground, in nests, and so on. Being given particularly only to mammals, doesn't such a perfect organ as the placenta reveal a manifestation of divine mercy and grace?

The fundamental difference between the metabolisms of animals in these two classes is also a big problem by itself. Mammals, in that they are warm-blooded, have every aspect of their lifestyle programmed accordingly. The body heat of mammals is kept constant by means of activating the heat-regulating systems in the hypothalamus region of their brains so that they can adjust to temperature variations. In contrast, reptiles are cold-blooded and their activities and metabolisms change with respect to the ambient temperature of their environment. We are unable to calculate how many well-directed mutations would be necessary in order for either type of metabolism to transform into the other. Beyond this, since reptiles cannot fly, how the wings of bat, a flying mammal, could ever have developed from the arm of a lizard is a complete puzzle.

As a matter of fact, even though he is an evolutionist, paleontologist Roger Lewin, who himself could not bear these troubles with evolutionary theory, confesses his feelings in the following words: "The transition to the first mammal, which probably happened in just one or, at most, two lineages, is still an enigma."[65] The neo-Darwinist evolution theoretician, George Gaylord Simpson, similarly expresses his displeasure about these quandaries in evolutionary theory as follows: "The most puzzling event in the history of life on Earth is the change from the Mesozoic, the age of reptiles, to the age of mammals. It is as if the curtain were rung down suddenly on the stage where all the leading roles were taken by reptiles, especially dinosaurs, in great numbers and bewildering variety, and rose again immediately to reveal the same setting but an entirely new cast, a cast in which the dinosaurs do not appear at all, other reptiles are supernumeraries, and all the leading parts are played by mammals of sorts barely hinted at in preceding acts."[66] The noted zoologist, Mark Ridley, of Oxford University, also points to the dead end to which so many unresolved questions bring evolutionary theory: "In any case, no real evolutionist, whether gradualist or punctuationist, uses the fossil record as evidence in favor of the theory of evolution as opposed to special creation."[67]

The Horse Story

Almost every introductory biology textbook contains popular pictures of the purported evolution of the horse: images show the tiny *Eohippus* prancing through the glades; then getting larger, more sure-footed, and faster, as shown through another series of "artists' renderings"; and finally, looking like the thoroughbred of today. On a television show on PBS entitled, *"Did Darwin Get It Wrong?"* Norman Macbeth, a Darwinian scholar, finally exposed the great horse caper that had gone unchallenged for close to eighty years, when he stated that they are not a family tree referring to the exhibit at the American Museum of Natural History; they are just a collection of sizes. For him there are no phylogenies.[68]

The drawings and models thought to be representing the evolution of the horses have been frequent evidence for evolution, and these are shown to students in evolution classes everywhere. Yet even though he was an evolutionist, Boyce Rensberger expressed the fact that there is no such foundation for the evolution scenario in the fossil records—and discussion of the "process of the supposed gradual enlargement of horses," whereby they are theorized to have thus reached the size of today's horse, has never even occurred at any meeting where the problems of evolution have been discussed. As he stated at the Field Museum of Natural History in Chicago, in 1980:

> The popularly told example of horse evolution, suggesting a gradual sequence of changes from four-toed, or fox-like creatures, living nearly 50 million years ago, to today's much larger one-toe horse, has long been known to be wrong. Instead of gradual change, fossils of each intermediate species appear fully distinct, persist unchanged, and then become extinct.[69]

The well-known paleontologist Colin Patterson, a director of the Natural History Museum in London, where the "evolution of the horse" diagrams were on public display at that time, said the following about the exhibition:

> There have been an awful lot of stories, some more imaginative than others, about what the nature of that history of life really is. The most famous example, still on exhibit downstairs, is the exhibit on horse evolution prepared perhaps fifty years ago. That has been presented as the literal truth in textbook after textbook. Now I think that is lamentable, particularly when the people who propose those kinds of stories may themselves be aware of the speculative nature of some of that stuff.[70]

In sum, this scenario was founded on deceitful diagrams and models devised to present the sequential arrangement of fossils of distinct species—which lived during vastly different periods in India, South Africa, North America, and Europe—solely in accordance with the rich power of the evolutionists' imaginations. More than twenty such charts have been proposed by various studies, each presuming to depict the evolution of the horse, though each is totally different from the other. Therefore, it is obvious that evolutionists have been unable to reach common agreement on these so-called "family trees." The only common feature in these arrangements is the belief that a dog-sized creature called *Eohippus (Hyracotherium),* which lived during the Eocene period, 55 MYA, was the ancestor of the horse. However, the fact is that *Eohippus,* which became extinct millions of years ago, is almost identical to the *Hyrax,* a small rabbit-like animal which still lives in Africa and has nothing whatsoever to do with the horse. Indeed, the inconsistency of the theory of the evolution of the horse becomes increasingly apparent as more fossil findings are gathered. Remnants of modern horse species (such as *Equus nevadensis* and *Equus occidentalis*) have been discovered in the same fossil layer as *Eohippus.* This is an indication that the modern horse and its so-called "ancestor" actually lived at the same time, and the evolution of the horse has never occurred at all.

The evolutionist science writer, Gordon R. Taylor, who died in 1981, explains this little-acknowledged truth in his book, *The Great Evolution Mystery,* which was published posthumously:

> But perhaps the most serious weakness of Darwinism is the failure of paleontologists to find convincing phylogenies or sequences of organisms demonstrating major evolutionary change... The horse is often cited as the only fully worked-out example. But the fact is that the line from Eohippus to Equus is very erratic. It is alleged to show a continual increase in size, but the truth is that some variants were smaller than Eohippus, not larger. Specimens from different sources can be brought together in a convincing-looking sequence, but there is no evidence that they were actually arranged in this order in time.[71]

In fact, American paleontologists, Charles Marsh and Thomas Huxley, were the ones who designed the series which is now generally thought to demonstrate a sequence of horse fossils as evidence for evolution. They arranged the sequence of horses—*Eohippus, Orohippus, Miohippus,* and *Hipparion*—with respect to the number of toes on both the front and back feet, and the dental structures of the fossils which were claimed to have hooves. They added the modern horse (*Equus*) to their series and announced to the general public that the diagram they had made up depicted the evolution of the horse. According to his scenario, Marsh deliberately put the fossils in such an order that the size would reach that of the modern horse. However, he dismissed many inconsistencies and logical fallacies while contriving the series. According to Professor Garret Hardin, as more fossils were uncovered, the chain splayed out like a branched tree opposing the previously sequenced series. For sometimes short horses, and sometimes tall horses, indeed had appeared diversely.

Most importantly, even though he had found many horse fossils, evolutionary paleontologist, George Simpson, complained about the nonexistence of mounted skeletons of horse fossils in his book, *Horses,* saying the following: "As far as I know, there are no mounted skeletons anywhere of *Epihippus, Archaeohippus, Megahippus, Stylohipparion, Nannippus, Calippus, Onohippidium or Parahipparion*, and none in the United States of *Anchitherium* or *Hipparion*." [72] The following observations by David Raup are also enlightening:

> The record of evolution is still surprisingly jerky and, ironically, we have even fewer examples of evolutionary transition than we had in Darwin's time. By this I mean that some of the classic cases of Darwinian change in the fossil record, such as the evolution of the horse in North America, have had to be discarded or modified as a result of more detailed information—what appeared to be a nice simple progression when relatively few data were available now appears to be much more complex and much less gradualistic. [73]

Numerous fossils have been examined up to this point, in terms of either the number of teeth, or toes, or vertebrae; as a result, it has been shown that the imaginary horse evolution scenario consisted of a great many inconsistencies. Further, such a scenario is always certain to be rejected if the different animals that were living in the past and are now extinct are sequenced simply with regard to a specific ideological orientation or prejudgment. (Note that the answer to the specific claim that horse toes somehow "became dull" to create hooves will be given later in the question related to vestigial organs.)

Climbing Up the Stairs or Taking the Elevator?

Since the "phyletic gradualism" model, or evolution by increments, so to speak, requires separate evidence at each step, it has terribly burdened evolutionists; thus, the invalidity of such gradual improvements has eventually been understood. An alternative scenario of "punctuated equilibrium," having many dilemmas and deficiencies of its own, has simply been put in its place. Indeed, in terms of some of its aspects, it is more difficult to accept.

According to the concept of "punctuated equilibrium," new organism types appear suddenly. This is actually an evident escape from the present problem of lack of fossil evidence for successive changes—or a kind of bypass, in effect. The more insistently the presence of important punctuations during evolution is claimed, the less need there will be for intermediate forms. For his part, Darwin dedicated himself to clarifying the mysterious absence of the innumerable intermediate forms, which are necessitated by gradual evolution, since he was categorically and unhesitatingly against the idea of punctuated evolution. Right before the publication of *The Origin of Species*, in fact, Thomas Henry Huxley (1825–1895) wrote the following in his letter to Darwin, dated November 23, 1859: "You have loaded yourself with an unnecessary difficulty in adopting *Natura non facit saltum* [Nature does not make leaps] so unreservedly."[74]

The inclination to see evolution from a punctuated point of view is founded on the "punctuated speciation" model articulated by American paleontologists, Niles Eldredge and Stephen Jay Gould. They accepted the gaps as natural phenomena; in fact, they considered them to be the result of the evolutionary mechanism, rather than assigning them to the shortfalls in the fossil records. According to the model of punctuated evolution which they offered, the development of living being was a process which occurred in stages by means of certain long, discontinuous, static periods. New species within a group around small isolated populations, for instance, appear very rapidly. The changes leading to a new species do not usually occur in the main population of an organism, where changes would not endure because of much interbreeding among like creatures. Rather, speciation is more likely to happen at the edge of a population, where a small group can easily become separated geographically from the main body and undergo very rapid morphological changes that can create a survival advantage and thus produce a new, but non-interbreeding species due to mutations. Having breeding capacity, that small number of species was then understood to be transforming into a new species. Yet since the non-interbreeding species were not able to spread widely, their fossils could not be found. So how about the presumed-to-be thousands, and even millions, of intermediate species? Were all of those species assumed to be "non-interbreeding small populations" within their isolated regions? Is such an assumption even tenable?

The hypothesis of "punctuated equilibrium" has substantially been a staged media event. It was specifically developed to try to account for the nonexistence of intermediate varieties between species—but as a kind of ironic twist, its main influence was to take public attention right to the gaps in the fossil records. As a major result of the appearance of the theory of Eldredge and Gould and the media campaign, for the very first time, the community of biologists clearly and consciously realized the absolute nonexistence of transitional forms. After the unfolding of "the trade secret of paleontology," in Gould's words, the old comforting belief that fossils would someday provide evidence of evolution through gradual changes weakened so much that it made backtracking impossible.

In fact, paleontological evidence does not offer any convincing proof that could make us believe the evolutionary model—which argues for continuous change in life forms and leaves the gaps between forms completely unexplained. A couple of species or groups which seem to be intermediate forms, at least to some extent, like *Archaeopteryx* or *Rhipidistian* fishes, might be brought to mind. Yet even though these fishes do have certain properties with regard to some distinct aspects, there is no corroboration that they carry characteristics of intermediate forms any more than some of the groups living today, like dipnoi (lungfish) or monotremes (single-cloaca mammals). However, those living groups which are characterized as being "intermediate forms" are certainly and obviously isolated from the groups that are claimed to be their closest relatives, and they do not embody transitional organ systems. Furthermore, it is very hard to even imagine a transition in any organ—for example, one simply cannot envisage the shift in respiratory organs between lunged and gilled fish; and there is not a single shred of evidence in existence about how the transition from a monotreme's distinct excretion and urinogenital system to that of mammals would have occurred.

Let us start with an analogy and let us imagine the vertebrate classes as private apartments in a five-storey building. Fish dwell on the first floor; amphibians are on the second floor; and reptiles, birds and mammals are on the third, fourth and fifth floors, respectively. Now, let us search for the possibility of the emergence of amphibians from the second floor to the level of reptiles on the third floor. There are actually two ways to go from one floor to another: you either take an elevator and ascend swiftly without getting tired, or you climb up the stairs one by one gradually. Now, if we move from analogy to reality, the idea of climbing up the stairs gradually represents "gradual evolution," while taking an elevator and going up swiftly represents "punctuated equilibrium." Further, let us consider that an amphibian on the very first step of the second floor stairs has 90% of amphibian properties and acquires 10% of reptilian properties, by means of a few random mutations. On the next step, it will effectively embody 80% of amphibian properties and 20% of reptilian properties, since its amphibian characteristics will be reduced and its reptilian characteristics will increase as it goes up the stairs, so to speak. Then, at the last step of the second-floor stairs, it will essentially display 10% of amphibian properties and 90% of reptilian properties, after which it will eventually reach the third floor and become a reptile.

The practical equivalent of such a hypothetical scenario is the existence of intermediate living forms belonging to each step, but that has never been the case; that is, the fossil record of even a single linking species has never been found. Further, the notion of such gradual changes has always faced serious difficulties due to the expectation of small, well-directed mutations, one followed by the other, in virtually every single organ and body system, while the animal "moves up" each one of the steps to the next. Seeing the impossibility of gradual improvement in the face of the dead-end of "no linking fossils," and of the absolute implausibility of such synchronous and "well-directed mutations," evolutionists have proposed the alternative, punctuated equilibrium, simply to allow themselves to claim that it is possible to jump from one step to another, or by taking an elevator, even to jump from one floor to the other.

However, this alternative is not as problem-free as it is claimed to be; rather, it is, in many ways, a worse dead-end than the previous model. This is because in such a case, in order for a fish to go through hundreds of changes and become an amphibian, we will have to overcome the impossibility of random occurrences of larger, well-directed mutations on the same individual, at the same time. Even if we assume the possible co-incidence of a couple of specific mutations, the changes that those mutations cause will result in defective body parts, that is, tissues and organs with deficiencies. We are unable, in fact, to calculate how many mutations, and how many millions of years, would be required in order for a fish's skin, covered with bony scales, to transform into a frog's bare skin, covered with poison glands. Besides, if we included in that calculation the transformation of fins to lungs, or the change in the heart from two chambers to three, we would not be able to reach any conclusion other than attributing infinite power and knowledge to those supposedly "random" mutations.

Classically, cladism is an evolutionist method of classification. Based strictly on the distinction of primitive characteristics and derived characteristics, it establishes a schema which presumes evolutionary relatedness among various groups of species. For example, since it is accepted that a lizard and Eurasian goat have common characteristics (of course, such a claim is necessarily initiated by such a presumption), their relationship is assumed to be close with respect to a common carp. Presuming a connection between these two species in the past, the presence of common ancestral links between them, termed phylogenies, is claimed. However, according to the advocates of evolutionary theory, this common ancestor between the lizard and goat is actually younger than the common ancestor of all three—the lizard, goat, and carp—and this seems like a complex technical problem at first glance; nevertheless it has to do with the fact that the cladistic approach is actually closely related with Marxism. Since it also means the denial of the gradual evolution of organisms over time, for some evolutionists, this approach has not only lost its acceptability scientifically, but it is also considered ideologically dangerous. That is why cladism, which is founded on discontinuous evolutionary thinking, is deemed to be incompatible with the teachings of Darwin and other theoretical pioneers in this area, like Ernst Mayr.

On this point, Popper expressed a determined view that Darwinian theory was not reliable enough—rather, it amounted to arbitrary speculation. As he saw it, so many issues about it were unresolved and another theory would be able to explain the same phenomena more comprehensively and persuasively: "I have come to the conclusion that the concept of evolution by natural selection is not a testable scientific theory, but a metaphysical research programme—a possible framework for testable theories."[75]

Dr. Beverly Halstead (1933-1991), from University of Reading in the United Kingdom, thought that human history could be analyzed in two ways: it could be explained according to either schema, which were based on a "gradualism" principle (whereby changes were gradual and not sudden) or a "revolutionary" principle (whereby changes were swift and there were "jumps" and discontinuities). Believing in gradual evolution himself, Halstead argued that the second type of evolution was Marxist in style, and this is something that was actually proposed by both Engels and Stalin. Accordingly, it is fundamental to admit that changes in acquired characteristics are not gradual, but there are swift and sudden jumps from one state to the other.[76] What he was obviously claiming was that if the occurrence of jumps could be admitted in the biological sciences in a way that would explain evolution, Marxist ideology would gain strength.

Unfortunately, presuming that they were learning science objectively, the British public were being misguided. According to Halstead, those who were in charge of the British Museum, for instance, introduced dinosaurs and human fossils with exaggerated respect for the classification method called "cladism" and misused their authority.

Nonetheless, this call was not very effective in convincing everyone. Harry Rothman pointed out that Marxists were not the only people who believed in discontinuities, and he asked the following question: "Will it be necessary to reject all scientific theories and explanations that apply to sudden changes from now on?" In this regard, for example, some might request the abandonment of the "Big Bang" theory. Was all this backlash against the notion of "sudden change" really essential?

Furthermore, was it even true that cladism included an interpretation of "intermittent" evolution? A great many biologists rejected this point. Cladism deals with systematics (taxonomy), but it does not offer any explanation of the rhythm or speed of evolution. To paleontologist Colin Patterson, from the British Museum, Halstead was confusing the existing problems with another. For Patterson classifying species is different from offering an explanation on how those species evolved. Besides, not only advocates of cladism supported a punctuated equilibrium. T. H. Huxley, who was among the ardent supporters of transmutation in the nineteenth century, was also an advocate of discontinuity, and he regretfully opposed Darwin's prejudgment that "nature does not jump." This was because the long and significant gaps in the history of life had showed themselves.

Being accepted as a way of classifying living beings, cladism does not give any reason for the speed or mechanisms of evolution, while its structure is contrary to the evolutionists' notion of a "common ancestor," and the implication that there is a "common ancestor" at the branching points of "cladograms" emphasizes the ideological extent of the problem. The most important evidence for this is that none of those drawings, which are made by returning millions of years back in time, can be either observed or tested.

Since the idea of cladism goes against gradual evolution, it was accepted that sudden changes must be possible in order to explain the existence of any species. Beyond cladism, a key remark on another theory, called the "sudden emergence of species," by S. J. Gould and N. Eldredge, was as follows: "The most important part of evolution does not occur in a local area, but it occurs in isolated small populations at distant regions as a rapid speciation."[77]

There was no doubt that such a theory was the perfect compromise with the idea of "discontinuities." While supporters of gradualism could claim that local micro-mutations were deposited gradually over time, advocates of "the sudden emergence of species" could claim that periods where no evolution happened were interrupted by "the emergence of new species." On the other hand, Gould was not able to explain how the new species would have arisen suddenly or swiftly, nor could he propose how such a mechanism could be interpreted scientifically. Instead, he proposed this explanation: "The chance of finding evidence for sudden emergence of the species is very weak since the change occurs in a very small population very rapidly."

Of course, these statements were not those expected from a scientist, for no one has ever observed in nature what he claimed, so it was merely an assumption. Yet weren't those who believe in creation also accepting the possibility of the occurrence of the very same thing—that is, that God suddenly created the species? Besides, the time span of God's creation is unknown to humankind. Therefore, wasn't the difference between the two views simply a matter of belief? Thus, shouldn't both views be given the same status in the context of subject matter where the chance of finding proof is weak? Put another way: is it really fair to accuse only those who believe in creation of being unscientific?

Although what they express is different from Darwin's view, Darwin used a similar method when mentioning the time duration of natural selection: "As natural selection acts solely by accumulating slight, successive, favorable variations, it can produce no great or sudden modification; it can act only by very short and slow steps."

In other words, whether it is gradual or sudden, evolution occurs in such small details that it cannot be seen by the human eye on fossil remains—and specifically how these changes could occur over a long period of time, during the entire life of an individual belonging to any species, is indeed a mysterious phenomenon that apparently never leaves a footprint behind. Because of this, there is no point in even rejecting such a "scientific" interpretation. In fact, it has been shown again and again that this is a problem which cannot be solved by the methods of science. As a matter of fact, the particular point which some do not like to understand is that a "theory" is nothing more than a model which is advanced in order to explain some phenomena, and it is always open to being disproved. In this case, it would be more accurate to call these opinions about evolution "hypotheses" rather than to consider them collectively as a "theory."

The British magazine, *The Guardian Weekly,* commented on Eldredge's interview with a group of science journalists as follows:

> If life had evolved into its wondrous profusion of creatures little by little, Dr. Eldredge argues, then one would expect to find fossils of transitional creatures which were a bit like what went before them and a bit like what came after. But no one has yet found any evidence of such transitional creatures. This oddity has been attributed to gaps in the fossil record, which gradualists expected to fill when rock strata of the proper age had been found. In the last decade, however, geologists have found rock layers of all divisions of the last 500 million years, but not a single transitional form was contained in them.[78]

As a matter of fact, the aim behind proposing this theory over those who explain the sudden emergence of new species on Earth with the idea of "creation" was not only to "claim" to be scientific, but also to try to explain processes which were neither observed nor pointed out by the "gradualism" of Darwin's evolutionary theory.

According to the theory of the sudden emergence of species, any type of species could be divided into a new subgroup, causing a new species to emerge within a relatively short period of time. Later on, following a more or less lengthy "balancing" or "stabilizing" period, a new subgroup would start to operate—and such a process was presumed to be continuously going on. So, where did this theory stand with respect to cladism and Darwinism? Is it really close to cladism?

According to Halstead, the answer was, "Yes." There was a certain relationship between the cladistic approach and the theory of "sudden emergence of species." In particular, Eldredge and Gould made use of this theory in a manner similar to that of Hannig, who was considered to be the father of cladism. Nonetheless, so many scientists found such assertions insufficient and baseless.

S.J. Gould sent a letter to *Nature* magazine stating that he was not a cladist. Further, in his letter, he explained that the theory of "sudden emergence of species" was itself dealing with the rhythm of evolution whereas cladism did not propose any explanations concerning this.

Link with Marxism

According to Halstead, the concept of "punctuated equilibrium" and Marxist ideology were based on the same philosophy; in other words, changes occurred by jumps in both. Gould has related how he learned about Marxism in his very early childhood. Although one of the founders of the theory of sudden emergence of species, Eldredge was nonetheless not a Marxist. Engel's book, *Dialectic of Nature*, along with many other books on the subject, undoubtedly contained interesting information. However, it was not easy to produce a complete and determined proposal which could define scientific thought as "dialectic." According to Halstead's interpretation, the pivotal concept was the notion of "jumping," and that was the contradiction between Darwinism and Marxism.

In trying to explain the classical evolution scenario that is based on gradual evolution and the concepts of Marxism, a geneticist from Cambridge, Gabriel Dover, pointed to an example provided by Engels: *"If water is continuously heated, there will be a gradual increase in its temperature; upon reaching the certain threshold value, it will start boiling."* In other words, there was a "jump" which could not be considered separate from gradual evolution. In biology, Darwin's theory also proposed the same scheme: *"Small quantitative changes accumulate and this process unavoidably causes a change in the true nature. In such a case, classical Darwinism is most compatible with Marxist theory."*[79]

Considering these claims, accusing cladists of being Marxist was at least controversial. According to Halstead, however, ideological factors also played a role. Indeed, there were influences between certain ideological concepts and scientific interpretations which were happening "undercover," so to speak. For instance, in their article where they proposed the theory of the "sudden emergence of species," in 1977, Gould and Eldredge clearly stated that gradualism was politically manipulated to accommodate the sociocultural tradition of Britain in Queen Victoria's time (1837-1901). This meant that Darwin considered evolution as a continuous process because of a certain philosophical and social conditioning. Because of that, he looked at nature from the point of view of a particular ideology; there was a continuous change, but it was in harmony and unity with the prevalent values of Victorian England. At this point, it is clearly seen how Gould and Eldredge expound a Marxist explanation. Indeed, in spite of its gradualist aspect, Marx too found Darwin's theory attractive because of "the presence of struggle among living beings in nature." This he found both appealing and dangerous because it evoked the social and economic competition in Britain much more.

On the other hand, according to Gould and Eldredge, the idea of a certain biological discontinuity seemed close to the dialectical ideas of Hegel, Marx and Engel. Referring to a work published during the Marxist-Leninist era of the Soviet Union, Gould and Eldredge argued that it was not surprising for Russian paleontologists like Ruzhentsev and Ovcharenko to propose an interpretation of the "partial formation of species." According to Gould, however, this similarity between theory and ideology should not be understood as the cause of their theory; in other words, it was unfair to criticize the theory of the "sudden emergence of species" just by referring to Marxist sources. On the other hand, it is impossible to deny the presence of the above-mentioned philosophical and political background considering the mutual interference between science and ideology. Could the observable "phenomena" not simply be tested, instead of dealing with the notions of Marx or Darwin?

Moreover, M.J. Hughes-Games of Bristol University stated that the evidence for gradualism was much weaker than Halstead thought—and Phillippe Janvier even concluded that the so-called evidence was the product of illusion. Unfortunately, the contradiction between ideas rooted in culture and ideology caused the matter to erupt into a battle which seemed to be "religious"—for while no criticism was made, the process of "excommunication" was allowed to operate. The excessive level of chaos raised the question: Since Neo-Darwinian theory is so fragile and open to debate, does it even deserve to be evaluated as a scientific theory?

Indeed, this matter was boldly expressed to the public by those in charge of the British Museum. Colin Patterson gave the title "*Is Evolutionary Theory Science?*" to one of the chapters in his book about evolution. For Patterson evolutionary theory is neither completely scientific like physics nor completely far from scientific aspects. According to Halstead, this judgment was nothing short of scandalous. He started to mount a heavy opposition in the journal, *New Scientist*. What would be the end of this story, if we were to believe those who claim that Darwinism is not truly "scientific"? Wouldn't such a case be advantageous for creationists? However, the subject matter of debate became degraded into an effort to simply get the opponent to back down, rather than a search for truth.

Is Similarity in Appearance Sufficient?

While asserting—as if it were proven—that humans evolved from chimpanzees (or from a common ancestor to chimpanzees and humans), evolutionary theory does not actually rely on scientific evidence, nor does it use the type of language that the scientific method necessitates in trying to base its thesis on fossil remains in order to determine that an evolutionary process was experienced. Furthermore, advocates of evolutionary theory have not been able to find what they have been expecting from the fossil record for one and a half centuries. As will be laid out in this part, the claim of the evolution of humans from apes does not have clear, supportive evidence nor is it methodologically "scientific"; at best, it can only be deemed an opinion or belief.

Summarizing the discussion briefly about the purported chain of ape-to-human fossils, the following mistakes and biased evaluations may be cited:

1. The fossils of apes which lived in the past and are extinct today are evaluated by sequencing them arbitrarily as transition forms between humans and apes. In addition to the big apes, like gorillas, which are still living today, there were smaller apes and hundreds of other primate species, such as lemurs, living in the past. Those ape skulls have been deliberately sequenced in a system which presumes to show gradual transition according to the scenario that evolutionists have imagined, so the impression of an actual transition from apes to humans has been created.

2. When the above-mentioned point is not convincing enough, they simply combine missing and defective bone pieces which are collected from different places. Then, they complete the missing parts with plastic material or plaster, again according to their imaginary scenarios, and mislead the public, as if humans were simply descendants of "one of the missing ancestral chains." Should the occasion arise, they can even completely fabricate fossils.

Many examples of misleading evaluations and fraud can be found. One of the best known of those false fossil constructions, known as "Piltdown man" (*Eoanthropus dawsoni*), pre-occupied the public for many years. This "fossil" was "found" by Charles Dawson near Piltdown, in England, in 1912, and it was determined to be 500,000 years old. It consisted of parts of a human-looking skull associated with an ape-like lower jaw. Many studies and projects revolved around it for more than forty years. In addition, 500 doctoral dissertations were written about Piltdown man. During his visit to the British Natural History Museum in 1935, the paleoanthropologist, H. F. Osborn, said: "Nature is full of surprises; this is one of the most important discoveries about the prehistoric times of humanity."

It wasn't until 1949, when the fossils were dated using the fluorine absorption technique, that the authenticity of the "discovery" was called into question. Kenneth Oakley, from the paleontology department of the British Museum, tested his new radioactive fluorine technique on the Piltdown man fossil in 1949 and proved that the jaw bone did not contain any fluorine. This result clearly demonstrated that the jaw had been underground for not more than a couple of years. Later, from other studies performed with this method, it was established that the age of the skull was only a couple of thousand years. Further, in 1953, Joseph Weiner, an Oxford professor of physical anthropology, discovered that the jaw had been deliberately given a unique wear pattern and purposely changed to fit the "Piltdown Man." A group of scientists, including Weiner and Oakley, then undertook new chemical analyses, including an improved fluorine test, and found that the jaw and teeth were not the same age as the skull and jaw—and that, in fact, they were not even fossils. The skull belonged to a 500-year-old human being, and the jawbone was that of a recently deceased orangutan! The joints were rasped and the teeth were added and arranged later on specifically to make it look human. All of the bone fragments had simply been artificially stained with potassium dichromate in order to make them look ancient. When the bones were dipped into acid, all the stains on the bones disappeared. Weiner, Oakley, and Oxford anthropologist, Wilfrid Le Gros Clark, were now certain that the Piltdown fossil collection was a fake—a hoax, in fact. Being one of the discoverers of the infamous hoax, Le Gros Clark expressed his wonder as follows: "The evidence of artificial abrasion immediately sprang to the eye, indeed so obvious did they [the scratches] seem it may well be asked—how was it that they had escaped notice before?"[80] It is an understatement to say that the revelation of the forgery of the Piltdown man fossil gave evolutionists a headache for a very long time.

An extensive scientific debate then began surrounding the reconstruction of another fossil from a pig tooth—"Nebraska man." Some interpreted this tooth as belonging to *Pithecanthropus erectus*, while others thought of it as belonging to *Hesperopithecus haroldcooki*. The reconstruction of such a fossil solely from a pig tooth actually became quite comical. This was because the evolutionists who fabricated a primitive evolutionary ape-man fossil from one single tooth could not stop themselves—instead, they even placed his wife right next to him. The problem started in 1922, when Henry Fairfield Osborn, the director of the American Museum of Natural History, declared that he had found a fossil molar tooth from the Pliocene epoch in western Nebraska. This tooth allegedly bore common characteristics of both man and ape. Great scientific arguments revolved around it, and reconstructions of Nebraska man's head and body were drawn based on this single tooth. Moreover, Nebraska man was even pictured along with his wife and children, as a whole family in a natural setting. Evolutionist circles placed so much faith in this "nonexistent man" that when a researcher, William Bryan, opposed such biased conclusions for relying on a single tooth, he was almost lynched academically. Nonetheless, other parts of the skeleton were also discovered in 1927, and it was then realized that the tooth actually belonged to an extinct species of wild American pig. Suddenly, all the drawings of *Hesperopithecus haroldcooki* and his "family" were hurriedly removed from evolutionary literature.[81]

Even in the best-case scenarios, skulls of "transitional forms" were completed based only on a couple of skull fragments, which were simply invented with prolific imaginations and then made to look very distinct and "realistic" in the hands of different artists. Different people, for instance, were able to construct fossils with different brain volumes from the very same skull material. Then, they simply engaged in extensive discussions about which of those fabricated fossils were more legitimate as evidence. As a result, the elusive basis on which evolutionary theory was constructed was shaken once again, and the picture became even more confusing and complicated.

In addition to fossil forgeries and the fossils of extinct apes, some fossils that evolutionists introduce unquestionably belong to real people. Fossils of humans who lived in different regions and various weather conditions include *Homo erectus*, *Homo ergaster*, *Homo heidelbergensis*, and *Homo sapiens neanderthalensis*. In the past, some human races which lived in the same period of time together might have crossbred, producing different "strains". The differences between these fossils, which are deemed to be subspecies (races) of the human species in terms of the taxonomic system, in fact are not any greater than the differences between Inuit people, Caucasians, African-Americans, Asians, or Australian Aborigines, for example—all of whom are currently living. However, evolutionists are determined to make such an effort to accept the idea of ancestral human races, like *H. sapiens neanderthalensis*—who was a rather stocky, extinct human strain—along with other human fossils as transitional forms. In another similar case, a skull and some bones of a purportedly human fossil named *Homo habilis* was eventually found to be, and reclassified as, an extinct ape.

One of the most significant difficulties in this field has been that once geologically dated fossils would not fit the evolutionary scenario after some time it was necessary to do changes on them.

Anatomical characteristics that were supposed to be seen only in modern man according to evolutionary schema, were observed in fossils from much earlier periods. Further, judgments were not made after analyzing a completely preserved skeleton of a certain living species. Rather, conclusions were exaggerated interpretations of studies of single bones—not even a complete bone would be taken into account, as only a fragment of it would be considered—from which they extrapolated deductions about the definitions of species.

In fact, the subject of how one can distinguish any type of species from another species by relying on such criteria as those now used is still open to discussion. Any human limb, or any part of a limb, could resemble an anatomically similar living being's equivalent limb, or a part of that limb, in some aspects. How "scientific" is it to take this similarity as a fundamental criterion, thus supposing that it gives an accurate result for determining species—and even for determining classes—rather than properly using it only as a base for scientific predictions or thoughts, and as a tool to open doors to new studies?

It is not satisfactory to "insert" into the scheme a living being which lived in the past and whose entire skeleton has not been found, only by relying on one criterion within any type of class. According to evolutionists, the samples of older specimens seemed more evolved than those of the more recent ones—for example, even though the teeth looked like hominid teeth, the jaw was totally an ape jaw. Besides, living organisms do not evolve in every aspect over time, as evolutionists idealize. Some of the organs remain unchanged, like those of the very old species, and some of them look like those of the most recent. So, in that case, which organ should be used in establishing the evolutionary relationship among species? Fundamentally, evolutionists get confused because of their elaborate and constraining preconceptions. Therefore, the same question should be asked here again: "Does evolution, which has never been proven, have to occur?" Why is it that they run away from explaining all these by reference to the ease and reasoning of creation?

The biggest error of those who do not accept a materialistic evolutionary perception has come from using the expressions of understanding that predominate in public opinion through the mass media. The main goal underlying the use of phrases derived from the classification and naming of animals according to the principles of systematic zoology for describing humans is to imply that humans are included in the same category as animals in evolutionary ideology. The notion of *primates* is such a powerful expression that it generates a completely artificial background, like other systematic categories, with the aim of examining about 600 ape-like species as an order bearing some common characteristics. On the other hand, one of the basic features of any animal taxonomic system is that it continuously changes with new discoveries. Included in rodents now, for example, an animal might later be included in a totally different group after a few years due to a recently found and distinctive feature it may have. Accepting that all the lemurs, tarsiers, lorises, chimpanzees, gorillas and orangutans are primates does not indicate that they came from a common ancestor—it only makes the researcher's studies easier. When the general characteristics of orders and families are known, it is possible to obtain typical information about the group without examining all the species included in the group one by one. However, evolutionists expel systematic zoology from its true orbit and put it into the service of the materialist point of view. In this respect, incorporating humans into the *Hominidae* family, with the name *Homo sapiens,* they placed the belief which belongs to their imaginary worlds into all of the zoology books, as if it were a reality.

Even though systematic zoology is a very important field which makes studying animal life easier and allows us to contemplate the beauties of creation, the ideological views of evolutionists have made many systematicians feel estranged. Since they have not been able to find any other way out, they have had to incline towards general acceptance and admit the imposition of considering human beings in the systematic categories of animals. However, humans are not living beings which can be evaluated only with regard to physiological or anatomical characteristics; rather, having intellect, consciousness and conscience, they are creatures which are completely distinct from animals in their essential nature. Thus, they should not be considered in these categories. Just as we divide vegetation, animals and bacteria from each other into separate kingdoms due to the differences in their natures, it has long been understood that humankind should be considered to belong to a distinct kingdom.

The answer to everyone's question about the distinctive characteristics of a hominid as opposed to other primates has never been provided. The three species examined within the family of *Hominidae* are the gorilla, chimpanzee and orangutan. The fourth species that the advocates of evolutionary theory include in this family is the human being. The distinctive characteristics of the other ape species included in the Primate order and those three species do not differ in their true nature. However, each species has unique features in terms of its morphology and anatomy, in addition to each having specific characteristics belonging only to itself. So should the human being not be distinguished from those species, in terms of both its true nature, and its "rank" or degree?

Contrarily, the subject matter is discussed in the domain of public opinion as if all of the problems had been overcome and an accurate result had been obtained. Even we were to look at things from the evolutionary point of view, and we were to accept such a taxonomic system, we would have to acknowledge that there is not a single bipedal primate alive today other than the human that permanently stands erect. Further, there is no other living primate with such a large brain to body mass ratio other than *Homo sapiens*, as the advocates of evolutionary theory call humans. If we look at the closest animals to us which are currently living—that is, to apes—we will see that they are as distinct from each other as they are from human beings. Similarly, none of the hominid fossils actually looks like a relative to humans. So, based on which criteria are those fossils "inserted" in this or that species, and then generally accepted by the public?

Most difficulties in paleoanthropology arise with the discovery of new, different and unpredictable fossils. The first of these "problem types" deals with the borderline between being ape and being human. However, the following reasonable evaluation could be achieved with a way of thinking that is sound and free of prejudgment: a human being is a "whole" and can only survive as a whole with a human identity. As a case in point, increased brain size and full bipedalism are given only to human beings, so that this wholeness is evidenced as being solely, and specially, for them.

The second problem faced while searching for the origin of humankind is the unwillingness of most paleontologists to learn about the variations in the fossil records—or even more basically, the inadequacy of their efforts. This indicates the apathy of many paleontologists, who opt to ignore this "scientific" problem, thereby using an approach which is inconsistent with the ethics of science—despite clear existence of variations in the fossil samples which have been put in the "human" category. The critical relevance of the question, "According to which criteria is it being considered human?", along with difficulty of solving of the problem with certainty, is simply ignored. The general estimation that something is "neither totally ape, nor totally human" arises from the attempt to represent populations belonging to a certain species with an insufficient and disorganized selection of fossils, in addition to the anxiety which results when attempting to define "humans" by referring to the science of biology alone.

The harmony of the human body, with its soul and essence, and the demonstration of the artistry of such wholeness on the face of the Earth, should make us think about the following points. The anatomy and physiology of our body, as given to us, is ideal for the purpose of the existence of our spirit, soul, mind, intelligence and senses; thus, we cannot call a living being, which partially has the features of humans but never displays those other characteristics which make it human, a "human being," because the resemblance is only partial. In other words, when a living being is said to be "human," it must typically have all of the characteristics which are present in humans altogether at the same time—not just a couple of features. Thus, it is only a human being if it possesses the following, and many more, characteristics at the same time: a greater brain to body mass ratio than the other primates; full bipedalism; a straight backbone and legs; compatibility of the length of the arms with the body and with the particular living conditions of man; a forehead projecting toward the front more than other primates; the ability to speak; intellect, conscience and reasoning; ethics, thus allowing it to be the interlocutor of revelation and religion, prompting it to bury the deceased, and permitting it to engineer complex devices, and so on; as well as many other characteristics which may or may not be reflected in fossils.

Yet, in terms of "representative types," evolutionists took only a single jaw fossil into account, and then they described the species by considering only this fossil. However, in the field of *systematic zoology*, a species is ideally described by a representative (*holotype*), which represents the species at its best, that is, the mature phase of ontological development. The question is, though, what is a sufficient characteristic to define humans? For example, since a human is not a creature that lives on trees, it is normal for the big toe to be adjacent to the other toes. Is this enough of a criteria for differentiation? At this point, the importance of gathering all the characteristics which make humans "human," and form an integral whole, becomes prominent once more as being the most critical requirement for describing humans. This is because humankind is such a complex creature, and we understand, again, that we have to evaluate it with all its characteristics—not by taking each minor feature, one by one, and comparing it to other creatures.

Are we human beings differentiated from apes or ape-like animals by our teeth? If so, then is it the shape of the teeth, or the enamel on the teeth, which is more important? Or is the clue about being a hominid present in the skull? Or is the joining of the back-bone with the base of the skull the distinctive factor? Or is it the shape of the elbow joints? Or the position of the big toe? Or all of these characteristics? Or, does the answer lie in another feature that is not considered above? Paleoanthropologists have tried to find the answer to the question, "What does being a 'hominid' mean?" Comparative anatomy experts, who have approached the subject matter ideologically, have discovered fossils that were claimed to be relatives of humans after determining the properties of human beings which apparently made them distinct from animals. Then, they evaluated these as if hominid fossils had evolutionary continuity from ape-like creatures to human-like ones. Furthermore, when the age and especially morphology of a fossil was not adequate for validating their anticipated results, they simply and abruptly changed their way of interpreting the fossils, and then continued to assert that they were hominids.

Ultimately, it is obvious that fossils do not provide an opportunity for evolutionists to talk about the status of human beings in the past. This incapacity is already present in the very nature of paleontology. Even so, upon finding a small bone fragment, a paleontologist or paleoanthropologist who has already espoused evolutionary thought assumes a right to base a very significant judgment on that very minor piece of bone.

Different human races have various skull shapes, forehead projections, nasal cavities, cheekbones, pelvis and knee joints, shoulder widths, different ratios of the length of arm and leg to the body, and so on—all of which are special to themselves and which are reflected in their fossils, even though these are admittedly missing and disorganized. For in terms of taxonomic systems, distinct human races are only different subspecies or varieties; put another way, according to Mayr's definition of species today, all human beings are from the same "species." That's because all human races can intermarry to produce fertile generations. As a matter of fact, differences can be observed in the shapes of the skull (and other morphological characteristics) even within individual societies in any region of the world. Therefore, this indicates that different geography, latitude, climates, eating habits, choices, and so on, can cause certain differentiations (as part of the genetic potential given to humankind during its first creation, and part of the natural range and limits of the "human" species). Indeed, in his book entitled, *Mankind Evolving*, the well-known geneticist, Theodosius Dobzhansky, reduced the case that taxonomists define as *variation* only to the level of variation among individuals of the same species (just like the formation of human races).[82] As a believer in evolution, Dobzhansky accepted that new arrangements occurring naturally on chromosome pieces allowed the idea of the emergence of new species, but after his experiments on fruit flies, he did not accept that human beings could have arisen as a result of such changes, like other organisms.

What makes evolutionists confused and always keeps them bewildered in human-ape debate is a problem caused by the nature of paleoanthropology itself, as news about the discovery of new fossil remains may come from any part of the world. After the age dating and morphological description of such a new fossil is completed, there is an attempt to place it somewhere in the current taxonomic systems. However, that usually shakes the arguments accepted thus far and necessitates "retouching" those theses. Examining related publications, the reader can observe that the date, place and form of the purported "split" between humans and apes and their supposedly common ancestors (according to evolutionary theory) changes from month to month, and year to year. Therefore, as we read above, evolutionists necessarily keep discussing what "portion of the criteria" described in their "theory" should be applied to the recently found fossil.

Nonetheless, the "movie" scenario described by paleontology and paleoanthropology has never been rewound to be viewed again. Studying in the face of so many obstacles, it should become obvious how difficult a job it is—and how much responsibility is required— to make judgments about the true history of human lineage.

In addition to that, humans are a presently living species. Therefore, comparisons between fossils and living forms give us a chance to make sound adjustments and establish standardization. Yet, if the human species were extinct, would we still gather the people of different races under the same species (as different subspecies), or under different species (that is, including them in different classes), just by looking at their fossils? Clearly, it is not even methodologically possible to say that there is an evolutionary relationship involving a transition from species to species between morphologically similar groups of living beings which lived in the past, and are distinguished from other species, only by examining their fossils today.

For instance, with a preconception that humans and apes are definitely related, Bernard Wood and Alison Brooks, of the Department of Anthropology at George Washington University, mention in their article published in *Nature* magazine that they are now almost certain that modern humans and chimpanzees diverged from a common ancestor which was chimp-like, predominantly arboreal, and fruit-eating, between 5 and 8 MYA. Nonetheless, there is a huge three-million-year gap between five and eight MYA, and there is absolutely no evidence about how they diverged during this big time interval. Yet the authors do not consider this big gap to be a significant methodological issue while arriving at interpretations, since they already have certain preconceptions in their minds. They continue: "Although we would expect human fossils to be considerably more bipedal than (and, thus, readily distinguishable from) the ancestors of chimps, this may not be so. Instead we may have to rely on the size and shape of the canines, as well as relatively subtle indicators in the deciduous and permanent post-canine teeth, to sort the first humans from the earliest chimps."[83]

This is actually a confession that there is no objection to making essential judgments despite missing information, even though the evidence presented is clearly insufficient. Not a single bit of fossil evidence, providing any information whatsoever about bipedalism, is present among the fossils belonging to this time gap.

Accordingly, the authors discuss the human-chimpanzee relation-ship based solely on some canine teeth. Indeed, not only 5 MYA, but even going back only 130,000 years, the possibility of finding human fossils, and especially of discovering fully-protected skeletal remains, is reduced by degrees. Even within the understanding, or assumption, of evolution, it becomes pretty hard to say anything certain about the characteristics of only one species, aside from try-ing to establish a possible relation of affinity or derivation between the species.

It could be said that evolution is merely a manifestation of prejudice. When the entire ideology is based on lowering humans to the level of animals, the understanding of some similarities that are given to challenge, or test, our understanding of life on Earth—or as a simple necessity for living in the physical and chemical conditions on Earth—can easily be distorted. The imaginary pictures of apes that seem gradually to become human beings, drawn one after the other, are only generalizations arising from prejudgments. It is being proven by new studies every day that presenting fabricated hominid fossils—by starting out with the partial similarity of a couple of bone remains—is unscientific and not relevant to science at all.

What Do Molecular Biology and Genetics Say?

Should one ask the question, "What is the greatest obstacle which faces the evolutionary hypothesis today?" —the answer will be "molecular biology." The first reason for this is that as a necessity of its field of interest, molecular biology deals with molecules—which are at the micro- and nano-scale, at the "borderlines of life," so to speak. The reality of "irreducible complexity" precludes the possibility of coincidence operating at the molecular basis of bio-chemical processes and operations to yield the amazing order, har-mony, system and plan, which are obviously observed at the micro-level. We have learned that life is much more complex than we could ever have imagined even thirty years ago. For instance, considered as the simplest living beings in most evolutionary taxono-mies, bacteria have been observed to have delicate structures con-sisting of hundreds of organelles at the micro-level, essentially pre-cursors of bio-chemical motors—tiny but incredibly complex and perfect structures—within their flagellum, in order to help them with movement.

All the evolutionist interpretations, which are based on super-ficial similarities shown as proof of evolution, as organs are seen simply "from the outside," became meaningless in a moment, when new discoveries brought researchers face to face with the perfect operation of dazzling complexity at the molecular level. The artistic construction and organization of the cell organelles themselves, each being like a bio-chemical factory, clearly reveals infinite knowl-edge and power. If we were able to understand the refinement of even one perfect structure, such as a chloroplast on a single green leaf—the chlorophyll-rich organelle which synthesizes sugar as food for the plant—there would no longer be a starvation problem in the world. Surely, no rational person could claim that such "intelligent machinery," which produces sugar from carbon dioxide and water using sunlight, could ever have arisen by chance. Further, the emer-gence of respiratory enzymes and coenzymes on the membranes of the mitochondria, which work as energy stations, cannot be imag-ined to have happened by themselves. In addition, no one could honestly assign the arrangement of two specific sub-units of RNA molecules to achieve protein synthesis in a ribosome as transfer RNA and messenger RNA—and the unique synthesis of all the proteins in a certain living being—to a mechanism with a mindless and unconscious nature. No one could reasonably claim that ATP and creatine phosphate, which are placed at the foundation of the muscular systems of all living beings and in the message mechanism of nerves—a chemical which is necessary for the motion of the actin-myosin filaments in muscles—had arisen by chance.

Beyond this, the claim, when they were first discovered, that cells were covered with a simple membrane, and that such a mem-

brane had arisen by itself, was challenged by the discovery by molecular biologists of an incredibly fine membrane structure. Rather than
being primitive, the structure of the cell membrane, named the fluid
membrane model, was made up of three molecular layers. Today, no
one is able to categorize such a sophisticated structure as being
"primitive," or as having "arisen by itself"—for it still bears so many
mysteries, and it is highly organized, with many functional units. In
fact, key aspects of cellular functioning are still not perfectly understood, such as the succession of glycolipids, phospholipids and glyco-
protein molecules, through a mechanism whereby they leave open
channels at certain points; how this regulates the system to transport
matter in and out of the cellular or subcellular domain; how the
unknown molecules are recognized by the special receptors on the
cell membrane; and the mechanism of canceration.

We can partially understand the structure of the golgi device—
which functions in many cellular regulation processes, like the secre-
tion of enzymes and hormones—just by looking through an electron
microscope. In turn, each of the other structures—such as centrioles,
which become active during cell division, microtubules, constituting
the microtubular spindle fibers which are necessary for chromosome
separation, and many more cytoplasmic structures—exclaim in its
own tongue that such exquisite artistry can only be made by the
Creator, Who can achieve everything in such a perfect way.

Besides, being a "kingdom" all on its own inside the nucleus, the
DNA molecule, composed of two helical strands wherein the entire
life program of the cell is programmed by four simple proteins
(known as A,T,G,C), in such units called genes, opens the brand-new
horizon of molecular genetic as a distinct miracle. For the creation of
unique features in all living beings is a result of the characteristics of
DNA, which can be coded in infinitely many varieties in all living
beings, as a common language from worms to fishes, from mice to
eagles, or from flies to whales. In brief, DNA, a universal molecule,
is obvious evidence for infinite knowledge and power.

Because of all this, we can say that evolutionary thought has drowned in the sea of molecular biology. When we nonetheless keep hearing evolutionists claim that "molecular biology proves evolution," we are simply left speechless. At this point, I recommend that readers take a look at Michael J. Behe's well-known book *Darwin's Black Box* for the finest answer to this claim.[84]

It is quite normal and reasonable for those genes which code some basic, vital, biochemical processes to be common in all living beings, since all beings live on the same Earth. In other words, the common presence of some molecules in many living beings due to the necessity of certain critical life functions—like that of the cytochrome or hemoglobin molecules, which are vital for the biochemical mechanism of respiration—does not indicate that they differentiated from each other. Not only a fly and worm, for example but also a dog and human need to use oxygen to live on this Earth; thus, the use of similar molecules in biochemical processes related to respiration is, of course, normal and to be expected. Such an operation shows a unique Creator Who knows all of the needs of all creatures and provides these needs in an optimal way.

Contrary to many years of continuous propaganda in which humans were claimed to be 98.7% similar to chimpanzees, the article entitled, "Chimp Chromosome Creates Puzzles," in Volume 429 of the British journal *Nature*, explains that human and chimp genes are actually much less alike than had been thought. Important variations have been found on the sequences of chimp chromosome 22 and its "equivalent," human chromosome 21.[85] A general commentary on this in the article simply states, "The first detailed comparison study done reveals surprising differences between human and chimpanzee genes." In the same article, the following words of Dr. Jean Weissenbach, from France, also appear: "Chromosome 22 makes up only 1% of the genome, so in total there could be thousands of genes that significantly differ between humans and chimps."[86] Therefore, this result brings Darwin's theory to a major dead-end in terms of the origin of the human being.

Being an expert in the domain of the Prehistoric and Quaternary periods, paleoecologist Jean Chaline points out the inability of molecular biology to explain the "past": "Some biologists assume that humans and chimpanzees differentiated from a common ancestor based on the similarity between the two species' biochemistry and number of chromosomes. This hypothesis is founded on the following assumption: molecular and biochemical evolution is regulated by neutral mutations systematically. However, in 1979, M. Goodman, who examined the analysis of amino acid sequences, proved that molecular evolution certainly occurred randomly and was not systematic at all. Therefore, the falsity of the above statement has unfolded."[87] The critical question about humans and chimpanzees, which were placed on the same branches of the "evolutionary tree" as an *a priori* judgment —"When did they separate and differentiate?"—is defined merely as a "challenging question" by classical evolutionists. Paleontologist, Pierre Darlu, states the following on this particular subject: "A factor called mutation rate (the number of mutations in unit time) has been studied in order to be able to give an answer to this question. This ratio, which is very hard to calculate, requires a calibration based on paleontological data, carrying gaps and uncertainties. However, the rate itself might change from one gene to the other, and even from one nucleotide sequence to the other within one gene, and it might speed up or slow down over time. Even though the statistical models account for all these parameters, the results carry the risk of going into a major uncertainty."[88]

As has been seen from the explanations above, it is essentially impossible for the advocates of evolutionary theory to find a field like molecular biology or genetics, and embrace it as a "life preserver from the past," so to speak—after witnessing the insufficiency of paleontology—in order to support their argument for evolutionary theory, which is claimed to extend throughout all geological periods. The studies which can be performed in these fields, and which experts conduct on specific historical periods of time, are limited to analyzing the DNA samples of skin tissue of well-protected pharaoh corpses, for instance, to uncover the relationship among individuals in that lineage based on a number of mummies. Researchers cannot go further than determining the relationship within that genetic line through the analysis of mitochondrial DNA (i.e. the establishment of maternal lineage through the transmission of "DNA" in the eggs of successive generations) in samples taken from nonpetrified human and animal bones dated in ten-thousand-year increments, from 10,000 to 50,000 years ago.

Consequently, a member of the French Academy of Sciences, the famed zoologist, Jean Dorst, concludes that "One chromosome difference between humans and chimpanzees, which seem to be the closest to each other in terms of biochemistry and chromosome number, is not sufficient to explain the difference in the human being's establishment of civilization on Earth and the chimpanzee's continuous stay in the trees."[89]

The studies on human evolution have been well established on traditional Darwinian dogmas. The first one of those dogmas is that change by evolution indicates itself by means of imperceptible, infinitely small modifications. Obviously, such beliefs, which have been taken as the fundamental base for research about ancestral and intermediate forms and dominated paleanthopological studies up to now, are still predominant. But what if that is not really the way things happened at all? In fact, the necessity for an extremely long time period for changes to occur is one of the reasons why Darwin's ideas fell from favor.

In every subsequent edition of his book, *The Origin of Species*, Darwin required a longer period of time for the evolutionary process to be observed. However, the Earth was not old enough to allow this evolutionary scheme to occur. While trying to explain how a species presumably underwent transformations over time, this special evolutionary model could not propose any explanation for how life had become so richly varied. Indeed, Darwin was aware of this problem. Yet his only illustration, or admission, about this was one which he gave in his book, *The Origin of Species*, showing "dotted lines" to demonstrate how ancestors changed over time and branched into many species. Unfortunately, even though he named his book, *The Origin of Species*, Darwin was quite unable to explain how one species might have "split" into two or more species.

As extreme numbers were established through sophisticated calculations, in order to model the time needed for amino acids and proteins to come into existence strictly "by chance" in the first atmosphere, the degree to which the idea of evolution through successive, random changes was becoming ridiculous started to be understood. Compared to the length of time needed for molecules to be ordered as cell organelles, then cells, then tissues, then organs, by means of the programming of DNA and RNA codes by chance in a chaotic medium, the age of our Earth was calculated to be only as long as the time it takes to blink.

Considering only the so-called "split" of apes and humans from each other, and in terms only of the differentiation of the cortex— which is the apparent seat of the functions of thinking, reasoning and understanding—probability calculations proved that the time needed for the number of random mutations which would necessarily have to occur at the right time and place is vastly longer than the actual age of the Earth. In addition, human beings are human beings not only by virtue of their brains, but also by all their seen and "unseen" organs, senses, feelings and thoughts, being complex from head to toe. When similar calculations are done for the development of other anatomical and physiological characteristics, there is simply insufficient time according to the age of the Earth to allow the required, necessary random mutations to take place, even just for the differentiation of our thumb in terms of its capacity for movement. The only logical solution to this mathematical quandary would be to shorten the required time span—in other words, to assume that all of the millions of transitional living beings are somehow "ready," and that thousands of mutations somehow occur continuously on each of those living mechanisms. However, this would contradict evolutionists' previous claims, since they have long insisted that a fully functional protein, for example, might randomly arise somewhere among the trillions of molecules. Yet, at issue is not merely the improbability of the emergence of a single protein molecule, but the improbability of the emergence of an entirely new faculty on a human organ, perfect in all aspects. Simply, the age of the Earth does not actually allow for the possibility of such random changes.

As a similar claim to that of non-functional organs, advocates of evolutionary theory argue that most DNA sequences which are dysfunctional or useless, even though they were of value in the past, have "become junk" during the evolutionary process, over a long period of time. But as the Human Genome Project nears completion, the many hidden riches of so-called "junk DNA" have begun to be explained. For Evan Eichler, an evolutionist scientist of the Department of Genome Sciences at University of Washington the term "junk DNA" is nothing but a reflection of our ignorance.[90]

It is well known by now that information about protein synthesis, which is vital for our cells, is coded in the DNA in our genes. While the presence of 100,000 genes had been previously estimated for the human genome, researchers from the Human Genome Project have announced a new estimate of only about 30,000. The number continues to fluctuate, and it is now expected that it will take many years to agree on a precise value for the number of genes in the human genome. Only a very small portion of our DNA is coded as genes, and because of the fact that the rest of DNA does not contain instructions or codes for proteins, it is considered to be "non-coding DNA."

Some portions of non-coding DNA are accumulated between the genes, and they are referred to as "introns." Some of these non-coding DNA pieces form long chains in a way which repeats the same nucleotide sequence. Any highly complex and sequenced DNA piece (thereby resembling a gene) which has been found among those parts that we call "repeating DNA" has become known as a *"pseudogene*," and evolutionists have argued that these are nonfunctional gene segments which are left over from the evolutionary process. Since advocates of evolutionary theory are used to making such ascriptions, they eagerly name this genetic material "pseudo," "atrophied," or "junk"—without actually proving the nonfunctionality of these biological mechanisms. However, the fact that these "pseudogenes" are not being used in protein coding does not prove that they have no function whatsoever in any biological processes. In fact, the progress made in related studies in the past decade have proven such contentions to be empty illusions. As a result, these DNA segments are no longer described as junk—rather, they are characterized as "genome treasures."

As a matter of fact, even the observation that repeating DNA segments, which are in the heterochromatin regions of chromosomes, have no visible role in protein synthesis, should not earn them the designation of "junk" DNA. Yet since the subject matter is approached with prejudgment, such nomenclature is hurriedly given, and minds become confused. Renauld and Gasser, of the Swiss Institute for Experimental Cancer Research, state the following: "Despite its significant representation in the genome, (up to 15% in human cells and ~30% in flies), heterochromatin has often been considered as 'junk' DNA—that is, DNA without utility to the cell." However, they have found out that those DNA segments actually play a collective role in meiosis—cell division during reproduction.[91] Indeed, recent studies have proven that heterochromatin could play important functional roles. Individually, nonfunctional nucleotides become functional when they are gathered, or work, together. So, as Emile Zuckerkandl expressed it, "Despite all arguments made in the past in favor of considering heterochromatin as junk, many people active in the field no longer doubt that it plays functional roles. ...Just as, quite some time ago, populational thinking became a necessity in genetics, we need now to get used to populational thinking in regard to the function of nucleotides. They may individually be junk, and collectively, gold."[92]

In 1994, a molecular biologist, Michael Simons, at Harvard Medical School, in Boston, as well as a physicist, Rosario N. Mantegna, of Boston University, and some colleagues, applied two "linguistics" (sequencing) tests to genetic material from a variety of organisms that they assumed to be either simple or complex. That material was comprised of 37 DNA sequences, containing at least 50,000 base pairs each, as well as two shorter sequences and one with 2.2 million base pairs. Both coding and non-coding regions were represented in this material. In the end, they found structured "language properties," just like human languages, in this "non-coding" DNA, that is, in the 90% of the DNA which had long been ignored as "junk in the cell." As seen in all other dialects, the "language" was coded in such a complex and miraculous way that it could not possibly be explained as having happened, or been formed, by chance.[93]

It was discovered in another study that non-coding DNA in eukaryotic cells is actually a functional unit in the nucleus.[94] Researchers have observed a certain proportional relationship between the amount of non-coding DNA and the size of the nucleus, and they concluded that this is an indicator of the necessity of such DNA for a bigger nucleus structure. It was then shown in subsequent studies that these DNA segments are vital for the structure and functionality of the chromosome,[95] for they play a role in such mechanisms as regulating the appearance of genes during embryological development.[96] especially functioning in the development of photoreceptor cells[97] and the central nervous system.[98] All in all, these studies have proven that non-coding DNA plays a vital role in the regulation of embryological development.

In conclusion, it is no longer accepted that introns are "junk." Just as introns have been admitted to have vital functions in the cell, an important study has been performed on mice, indicating that pseudogenes are also functional. That study defined the pseudogene as a gene copy that does not produce a functional, full-length protein, and it pointed out that the biological roles of pseudogenes are still not well understood, despite determined efforts. It was described how the human genome contains up to 20,000 pseudogenes. Later on, the role of pseudogenes in the regulation of messenger-RNA stability was also reported. In fact, as a result of changing these genes genetically via a trans-gene insertion, polycystic kidneys and bone deformities were exhibited in the resulting mutant mice. All these findings demonstrate that pseudogenes are neither nonfunctional nor useless; rather, they are very crucial DNA segments which have integral functions in specific regulatory processes.[99] A study termed, "Not Junk After All," by Wojciech Makalowski, of Pennsylvania State University, describes how repeating DNA elements, called "Alu sequences" constitute more than 10% of the human genome. While they have not been observed to code for proteins directly, the study demonstrated how Alu sequences become inserted into the coding regions of genes, resulting in the formation of new proteins, and their important role was further established.[100]

Similarly, in their studies on zebrafish, Shannon Fisher and her colleagues at McKusick-Nathans Institute of Genetic Medicine, at Johns Hopkins University School of Medicine, have provided similar explanations of why the notion of "junk" DNA should lose its validity since such DNA plays various roles in regulatory mechanisms in the gene.[101]

First of all, the similarities between different living beings do not answer the basic question of biology, the question of how those unique and profoundly complex organs and systems of various living beings have arisen, and Darwinism cannot give an answer to this question. On the other hand, so many similarities among even very distant organisms, starting from the common point of their all being alive, can be considered. For instance, you can say that there is a resemblance between humans and bacteria in terms of the fact that they are both alive. Both of them have a specific shape, are able to reproduce, and use energy. You can also associate fish with bugs and humans as all three use oxygen, eat food with their mouths, and expel refuse via an anus—and such similarities can go on, and on. Yet, does seeing the similarity between living beings demonstrate that they have become differentiated from a common ancestor by chance? Or, rather, does it show that they are the work of art of a Power with infinite knowledge? A useful analogy is that we use the same construction materials—such as wood, sand, cement and glass—to build either a small hut or a huge house, a villa or a skyscraper. But just considering this similarity, no one would ever dare to claim that a skyscraper evolved from a hut by chance; if someone did, that person would clearly be the target of ridicule. Instead, everyone would agree that both the hut and a skyscraper were the work of art of an architect, or builder. Analogously, then, living beings which are made out of the same materials and which may even live in common conditions—in other words, which have some similarities—do not demonstrate by virtue of such similarities that they originate from a common ancestor. Extending the example above, if one constructs a dwelling for shelter, it will have a foundation and a roof— but the soundness of the house may vary depending on the quality of the foundation and the roof. Further, as beings live on Earth, it is naturally to be expected that they will have some basic metabolic processes and structures which are favorable to the particular living conditions on Earth. Moreover, we know that designers and engineers use many comparable pieces in different types of technical systems and products. For example, bolts, pins, screws, or cables are used in a variety of devices because they are ideal for a particular purpose—and yet, a machine which has the same or a similar cable as another, for example, cannot be said to have arisen by evolving from it.

Consequently, the main question is this: Can these types of similarities be associated with Darwin's theory? In fact, they cannot be associated with his theory because living beings which are supposed to be close relatives according to evolutionary theory are sometimes observed to be genetically very different, while those which are supposed to be relatively unrelated may have very similar organs or genes. For instance, the human eye and the octopus eye are almost identical in terms of their appearance from the outside.

But this does not mean that we are relatives of octopi, and as we descend deeper into the delicate structure of each eye, some very important distinctions draw our attention. While the photoreceptor cells on the retina of the eye of the octopus are placed in a position that is the nearest side by which direct light comes to the eye, the photoreceptor cells in the human eye are placed in a totally different position, so that they are distant from the incoming light, and they are covered with nerve cells and blood veins. Is it not more reasonable to accept that these two eyes are the manifestation of the infinite knowledge of One Creator, rather than considering them to have originated from a "common ancestor"? Does acknowledging this restrict development, research, or invention?

EMBRYOLOGY

After paleontology, comparative anatomy, physiology and molecular biology, the field of embryology has been the most appealing for those who like to support, and try to prove, evolutionary theory. Jeremy Rifkin approaches the subject matter by saying: "Many of the classic arguments that have been used to support evolutionary theory are like malicious gossip. Once in circulation, they feed on themselves. They multiply and expand until they are so pervasive that any attempt to challenge their veracity seems all but futile. Nowhere is this more in evidence than when we examine the field of evolutionary embryology."[102]

"Ontogeny" is a biological term used for the development of a living being from the embryo phase to maturity. "Phylogeny" (the development of lineage), which is used to try to explain evolutionary development (by the advocates of evolutionary theory), is considered to chronicle the evolution of species and their transformation into new species. Ernst Haeckel (1834-1919), a German biologist and philosopher, combined these two words and proclaimed to the world that "ontogeny recapitulates phylogeny," in his book, *Generalle Morphologie der Organismen* ("General Structures of Living

Mechanisms"), way back in 1866, and in his other text, *Natürliche Schöpfungsgeschichte* ("The Natural History of Creation"), in 1867.[103] Haeckel asserted that, "During the development of the embryo, it passes through all of the various stages of the evolutionary development of its ancestors. The embryo represents a moving picture of the entire evolutionary history of life on Earth. If one were to watch a human embryo develop, what would pass before the observer's eyes is every single transformation in the long evolutionary sojourn of life, from the emergence of the very first living cell onward." This idea, the opinion about the entire process of human evolution displaying itself in the different phases of embryonic life, was very appealing, and it caught the public imagination.

Haeckel's "theory" quickly became popularized, and it even came to be seen as evidence for evolutionary theory. Thus, while talking about evolutionary theory, people often mentioned Haeckel's version of events enthusiastically. In fact, the idea that *"ontogeny recapitulates phylogeny"* is still present in many books which function as an "Introduction to Biology." Although it was abandoned by its architects a long time ago, many instructors still teach the same fictitious story to their students as though it were true.

Now known as "biogenetic law," Haeckel's idea finds absolutely no supporters among dedicated biologists. Yet, after having been imposed for more than 130 years on the scientific community, and having been the object of derision for more than fifty years, such an idea is somehow still present in biology books as a result of varied *ideological reasons*. According to many researchers, "Biogenetic Law (i.e., Recapitulation Theory) is as dead as a doornail. In fact, even though it became outdated as a subject matter of scientific discussion in the 1920s, it was not taken out from textbooks until the 1950s."[104] Yet some still insist on keeping it in biology textbooks—even though it has been often expressed outright by specialists at scientific meetings that such a theory is "absolute nonsense."[105] According to Walter J. Bock, from the Department of Biological Sciences at Colombia University, "Biogenetic law has become so deeply rooted in biological thought that it cannot be rooted out in spite of its having been demonstrated to be wrong by numerous subsequent scholars."[106]

As a matter of fact, biogenetic "law" does not have sufficient merit to be called a "law," scientifically speaking. As to Haeckel's assertion, similar to the embryo of mammalian animals, birds and reptiles, a human embryo also has *"gill slits"* during a certain period of its embryological life. But these purported "gill slits" are presented by evolutionists as though they were evidence for an embryo's passage through fish, bird, and reptile stages, on its way to becoming a mammal. It is true that a series of small dents called "pharyngeal clefts" are observed during a certain stage of embryological development, and they do look little bit like the particular openings around the neck of a fish which function as gills. But this resemblance is merely external—affecting only their superficial appearance. We now know that pharyngeal clefts do not open to the throat and never have a breathing function in Earth's vertebrates. Instead of transforming into dents or gills, the upper fold eventually develops into the bottom part of the chin and the middle ear canals; the middle fold changes into the parathyroid glands; and the bottom fold becomes the thymus and endocrine glands.

Yet advocates of biogenetic "law" always display drawings of these "gill slits" to support their arguments—even though this line of thinking is respected by the pioneers of embryology. Gavin de Beer, the former director of the British Museum, and one of the world's distinguished embryologists, notes that the theory of recapitulation had its ardent supporters until recently. He remarks concisely on the tenacity with which people cling to such an obvious fallacy by saying, "The idea that ontogeny recapitulates phylogeny has the characteristics of a slogan in that it tends to be accepted uncritically and die hard."[107]

In turn, in one of his articles in *New Scientist* magazine, Roy Danson contends that the widespread and persistent acceptance of such a ridiculous concept says as much about the entire field of evolutionary biology as it does about Haeckel's particular contribution, and it brings the following question into the spotlight: "Can there be any other area of science in which a concept as intellectually barren as embryonic recapitulation could be used as evidence for a theory?"[108] In other words, claiming that it is impossible to distinguish between vertebrate embryos—such as those of a fish, a chicken, a rabbit and an ape—in the early stages of embryonic development actually demonstrates nothing except one's ignorance of embryology. That is why Darwin, who was not an embryologist, "took advantage of the ideas of Von Baer, who was a famous embryologist at that time, by distorting them. *Not believing in evolution, Von Bear criticized this distortion until he passed away in 1876.*"[109]

Ernst Haeckel vigorously expounded this idea of "embryonic recapitulation" at the beginning of the twentieth century. Then, lacking any evidence to support evolution, Haeckel expressly set out to manufacture data. He fraudulently changed the drawings made by other scientists, of human, chicken and fish embryos, in order to increase the resemblance between these and to hide the dissimilarities. Eventually, as mentioned, it was discovered that the structure displayed as the "gill" by Haeckel was essentially the developmental substrate of the upper fold of the middle ear canals, the parathyroid glands, and the thymus glands, in reality. So the figment of Haeckel's imagination began to unravel. Today, the entire scientific community admits this as one of the worst cases of scientific fraud. The folds argued to be "gill slits" in this evolutionary "story" disappear in progressive stages as vital structures for the animal's life from this part of the embryo. Further, the purported human "tail"—which was so named by Haeckel and his followers because it appeared earlier in embryonic development than the legs—was found to be the human backbone.

George Gaylord Simpson, one of the first supporters of evolutionary thought, described how unrealistic Haeckel's "theory" was in the following words: "Haeckel misstated the evolutionary principle involved. It is now firmly established that ontogeny does not recapitulate phylogeny."[110] Among Simpson's other statements, the following also calls our attention: "Haeckel called this the bioge-netic law, and the idea became popularly known as recapitulation. In fact Haeckel's strict law was soon shown to be incorrect. For instance, the early human embryo never has functioning gills like a fish, and never passes through stages that look like an adult reptile or ape."[111]

Another interesting aspect of Haeckel's forgeries was that in the drawings which purported to show that fish and human embryos resembled one another, he deliberately removed some organs from his drawings, or else added imaginary ones. Indeed, he has been criticized widely since his time because of his actions and assertions, but those approaching the subject matter strictly ideologically have chosen not to pay attention to these criticisms. Michael Richardson, an embryologist at St. George's Hospital Medical School, in London, pointed out Haeckel's misleading ideas in his studies by stating, "We are not the first to question the drawings. Haeckel's past accusers included W. His of Leipzig University, L. Rütimeyer of Basel University, and A. Brass, leader of the Keplerbund group of Protestant scientists. However, these critics did not give persuasive evidence in support of their arguments."[112] In turn, subsequent, detailed studies conducted by Richardson in 1997, 1998, 2001 and 2002[113, 114, 115] revealed how grievously Haeckel had distorted his drawings.[116] Thus, Richardson clearly established Haeckel's forgeries, using the serious criticisms of W. His, which had been ignored in the past,[117] as well as the ideas of Brass (106), the findings of Rütimeyer,[118] and modern knowledge of embryology.[119]

The September 5, 1997 issue of *Science* magazine formerly announced recapitulation theory to be nothing more than a supersti-tion, in an article titled, "Haeckel's Embryos: Fraud Rediscovered." After explaining all the contradictions pertaining to Haeckel's draw-ings, the article stated the following:

The impression they [Haeckel's drawings] give, that the embryos are exactly alike, is wrong, says Michael Richardson... So he and his colleagues did their own comparative study, reexamining and photographing embryos roughly matched by species and age with those Haeckel drew. Lo and behold, the embryos "often looked surprisingly different," Richardson reports in the August issue of Anatomy and Embryology. Not only did Haeckel add or omit features, Richardson and his colleagues report, but he also fudged the scale to exaggerate similarities among species, even when there were 10-fold differences in size. Haeckel further blurred differences by neglecting to name the species in most cases, as if one representative was accurate for an entire group of animals. In reality, Richardson and his colleagues note, "even closely related embryos such as those of fish vary quite a bit in their appearance and developmental pathway. It (Haeckel's concept) looks like it's turning out to be one of the most famous fakes in biology," Richardson concludes.[120]

Jane Oppenheimer, an embryologist and science historian, touches the subject as well: "It was a failing of Haeckel as a would-be scientist that his hand as an artist altered what he saw with what should have been the eye of a more accurate beholder. He was more than once, often justifiably, accused of scientific falsification, by Wilhelm His and by many others."[121]

The most striking aspect of "recapitulation" was Ernst Haeckel himself, a faker who falsified his drawings while he was alive in order to support the "theory" he advanced. When he was caught, the only defense he offered was that other evolutionists had committed similar offences:

"After this compromising confession of 'forgery' I should be obliged to consider myself condemned and annihilated if I had not the consolation of seeing side by side with me in the prisoner's dock hundreds of fellow—culprits, among them many of the most trusted observers and most esteemed biologists. The great majority of all the diagrams in the best biological textbooks, treatises and journals would incur in the same degree the charge of 'forgery,' for all of them are inexact, and are more or less doctored, schematized and constructed."[122]

After all the conclusions reached based on the references above, let us return to the domain of our modern knowledge of embryology. Considering the embryological developments of vertebrate classes in stages, each class has a very specific type of egg. Based on the particularities of the egg, the zygote acquires different types of blastula and gastrula stages in embryonic development by dividing distinctly in each group. As a result, each class has a very unique development period and developing organ, whose formation originates during the gastrulation (digestive) and neurulation (neural) development periods of a fetus. While lungs and legs are developing on a land vertebrate, gills and fins develop on a fish. Gills are formed ectodermally, while lungs are formed endodermally. There is not even the most minor indication of a gill ever developing in the pharynx regions of reptiles, birds or mammal embryos. As mentioned above, the ectodermal foldings in this region are the beginnings of some endocrine organs, the middle ear channel, the chin and some laryngeal cartilage—and the related organs are formed according to the genetic code for each.

Besides, the protective covering (*vitellin membrane, gelatin cover, amnion, chorion, vitellus sac, allantois* and *placenta*) of the egg and embryo of each class of living beings has a unique shape and characteristic which is specific to it. All these forms, which develop outside of the embryo itself, are obvious stamps of the miracle of creation in that they cannot be explained in any other way than by acknowledging the conscious preference of the One Who has infinite knowledge and power, and Who knows the particular difficulties and conditions that the embryo will experience.

Vestigial Organs?

The other famous tale closely related to biogenetic "law" is the idea of "vestigial organs." As the argument goes, animals sometimes have organs which appear not to be fully developed, or even nonfunctional; these are then surmised to be "leftovers"—vestigial remnants of inactive (unused) organs, or "relics" of organs or bodily components which are found in some purported ancestors—from the evolutionary process. This opinion has unfortunately become widespread. Once upon a time, biologists made up a list of 180 so-called vestigial organs in the human anatomy. However, numerous experiments done since then have proven that those so-called "vestigial" organs have crucial functions in the human body—they are not useless after all. For instance, it is very well known today that the appendix plays a very important role in fighting infections.

Perhaps the most important part of the human body which has been claimed to be a vestigial organ in humans is the coccyx. In fact, while this part of the human anatomy is scientifically not a "tailbone," as it is commonly called, supporters of evolution claim that it is a vestigial tail which was previously present in humans. However, R.L. Wysong points out that this organ is not vestigial in the least: "Far from being vestigial, these vertebrae serve as an important attachment site for the levator, ani and coccygeus muscles to the pelvic floor. These muscles have many functions, among which is the ability to support the pelvic organs. Without these muscles (and their sites of attachment) pelvic organs would prolapse, that is, drop out."[123]

A detailed examination of the coccyx unfolds the truly wondrous aspects of this bone. Let's take a look at the detailed explanations given by a contributing writer for *Sızıntı Magazine*, Dr. Aslan Mayda:

> The coccyx, commonly referred to as the tailbone, is the final segment of the human vertebral column, of 4–5 fused vertebrae (the coccygeal vertebrae) below the sacrum in a triangular shape. It is attached to the sacrum in a fibrocartilaginous joint, which permits limited movement between them. The anterior surface is slightly concave, and marked with three transverse grooves indicating the junctions of the different segments. It gives attachment to the anterior sacrococcygeal ligament and the levator ani, and it supports part of the rectum. The posterior surface is convex, marked by transverse grooves similar to those on the anterior surface, and it presents on either side a

linear row of tubercles, the rudimentary articular processes of the coccygeal vertebrae. According to those who believe in evolutionary theory, the coccyx has reached us today as a vestigial structure, a relic from our ape-like ancestors, and it has no function. Yet, should one analyze this bone in detail in terms of its anatomy and physiology, one will readily see how important a function this bone actually has.

Furthermore, the coccyx has two bulges. These bone bulges prevent sliding to the right or left during sitting. Reminding us of a work of art, complete with a geometric aesthetic in its anatomical appearance, four ligaments render stability when a person sits down on a firm surface while, together with the sacrum, these provide integrity and firmness.

...

We should also consider that the coccyx has a coccygeal artery for nourishing itself, a coccygeal vein for collecting venous blood, and a coccygeal nerve—all of which are favorable to its structure. In addition, it has a coccygeal bursa, coccygeal substance, glomus coccygeum, and Luschka glands that secrete a fluid which provides lubrication. Yet such a special anatomical structure does not form in the embryological "vestigial tailbone"; rather it takes shape in relation to the anatomical structure of its environment. For example, some people have an extra rib as a bulge from the seventh cervical bone, as an anomaly from birth, which is called the "cervical rib." This is seen in some people even though the rib is not normally present on the cervical bone. Since this extra bone does not have any arteries, veins or intercostal nerves, it is supplied by the main vascular systems and nerves of other anatomical structures around it. If the coccyx were actually vestigial, it would have no need for the sophisticated arteries, veins, nerves and glands which are unique to its anatomical structure. Another characteristic of anomalous bones is that they cause illnesses that can only be treated by operations (i.e., by removing the bone from the body). For instance, those with the anomaly of a cervical costa (congenital fusion) on their neck, mentioned above, experience arm pain, numbness of the arms, and lack of energy—and the complaints end upon the removal of this bone. However, the removal of coccyx causes serious problems with both birthing and defecation. Moreover, the coccyx has particular muscles and ligaments, called the coccygeus muscle, the sacrospinal muscle, the sacral tubercle, and the anococcygeal ligament. Together with these ligaments, a unique muscle, which is called the sphincter ani externus muscle, attaches to the edge of coccyx. This muscle holds the anus closed by encircling the anal canal, and it becomes opened in response to a person's effort during defecation. In turn, it takes supportive strength from the coccyx for its continuous contraction by means of the anococcygeal ligament. When a person sits down, in order to make the load lighter, the coccyx assumes a position inclined toward the front—thus, the heavy load of the body is suspended by the action of ligaments and muscles. Due to the particular attachment of its muscles, the coccyx also has a potential range of movement, especially during defecation. There is pressure at the back of the coccyx when a person is sitting down, but the coccyx reduces this pressure by moving the only hinge joint toward the front. Those muscles which are attached to the coccyx also support the base of the pelvic bone, and thus of the birth canal; further, they support the base of the large intestine, and other veins and nerves, as a protective cover.

So, the questions which must be asked is, "If the coccyx is considered vestigial—that is, if it were not deemed to have been created with a specific plan in mind—then where would these muscles and ligaments attach?"

In order for the muscles to be able to function, they have to be attached to the bones. If there is nowhere to attach to, so that it is just hanging in space, a muscle cannot get sufficient strength and thus cannot fully function; it will become contracted and weakened. For example, the anus muscle which functions to hold the anus closed is connected to the anococcygeal ligament. If the coccyx did not exist, these muscles would not function; therefore, the anus would be weakened by the pull of the muscle on the opposite side. Those patients whose coccyx has been removed complain about the weakening of anal contractions and the feeling that some kind of hard mass is stinging their anus. Should the coccyx be designated as being vestigial, then the attached muscles and ligaments should necessarily also be considered to be vestigial. Yet then one would have to ask, "While purportedly forming itself through evolution, could the "vestigial" coccyx bone have organized and brought along the other necessary "vestigial" structures—like its vein, nerve, gland, ligaments, muscles and joints?"

Another important aspect is that the joint between the coccyx and the rump bone is expandable. This joint's flexibility allows the opening to widen by 2–2.5 cm during the birthing process. If this were not so, either the baby would die due to waiting too long in such a narrow passage, or there would be grievous rips and tears caused in the womb and anus.

As a matter of fact, the anatomical structure of this joint is created with such delicate refinement that it does not easily allow any position for the baby during birth other than the typical presentation, whereby the head comes out first. The most important part of this mechanism is the coccyx. The soft tissues do not contract much, due to the moving of the joint between the coccyx and the rump bone to the rear. For this reason, the normal presentation (crown first), which precipitates the movement of the baby head first down the birth canal, is compelled.

In addition to all these, the concave shape of the coccyx also assists birthing by supporting the baby's presentation of the head first, in the crowning labor position. If this bone were not in this particular shape, the head would not be able to rotate back, so this position would not be possible; further, the head's circumference at its largest part would cause major complications and injury to the baby during childbirth, namely fractures, nerve injuries, and anoxia (lack of oxygen), which would damage the baby's brain and the other organs, causing essential malfunctions which would impact on its entire life. Therefore, calling the multi-functional coccyx an "extra" bone, or an "unnecessary, non-functional" bone, is not a reasonable conclusion for a rational mind. In effect, arguments making such claims are merely ideas proposed with prejudice, without examining the anatomy, physiology, pathology, biochemistry or biomechanics of any organ.[124]

Similarly, even though some organs—like the tonsils, appendix, epiphysis tissue, parathyroid, thymus, body hair, and wisdom teeth—have all been mentioned as "vestigial structures" in the past, evolutionists now seem to be exhausted and do not have much to say to prove their arguments based on the idea of vestigiality. Contrary to what they have claimed about the supposed "nonfunctional," "useless," or "vestigial" nature of these structures in humans or animals, new research methods and technology show that all organs are created in a particular form for a particular purpose. Pages of information, from hundreds of sources, could be added here about the excellent harmony and cooperation between the functions of these organs and miscellaneous body activities. However, it is sufficient to return to our consideration of the appendix, which was long assumed to be vestigial, and note the new findings about this complex organ: "Goblet cells in glands in the appendix secrete a mucous lubricant into the intestines which aids the movement of material through them. After the appendix is removed, the patient suffers constipation and the risk of getting intestinal cancer increases."[125] Other recent findings about the appendix reach the same conclusion: "It is rich in lymphoid tissue, meaning that it acts as a filter and removes bacteria and protects the intestines from infection. A study done on hundreds of patients with leukemia, Hodgkin's lymphoma, Burkitt's lymphoma, cancer of the colon, and cancer of the ovaries showed that 84% of these patients had their appendix removed, while in a healthy control group only 25% had it removed."[126]

It has also been shown through modern immunological techniques that tonsils and adenoids are very crucial lymphoid organs for the immune system which not only produce antibodies but also function in cell-mediated immunity.[127] Likewise, it has been found that Hodgkin's lymphoma is observed three times more frequently in those whose tonsils have been removed.[128] The importance for the immune system of T-lymphocytes, which the thymus produces, has also been shown in recent studies. Further, being secretions of the epiphysis, which is sensitive to light, both melatonin and dimethyltryptamine (DMT) have been found to play a role in the regulation of sleep and the biological clock, and these have been shown to have other effects on the immune system and some endocrine glands, thereby affecting the reproductive season of animals, including patterns like hibernation—all of which emphasizes the importance of these "vestigial" structures for bodily health.

HOMOLOGY OR A COMMON PLAN IN CREATION

Another claim which is proposed as evidence for evolution is related to the interpretation of similarities. Certain types of morphological parallels are very common in nature: for example, the resemblance of the bony structure of the fins of the whales and the ichthyosaur; the resemblance of the eye structure of vertebrates and cephalopods; and that of the inner ear structures of birds and mammals. Even though all these similarities are very striking, there is not even the slightest biological affinity among those species in terms of their genetic program.

Based on a complete lack of evidence, therefore, homology is a superficial, imaginative notion which is proposed as a result of surveying the external appearance of things. To date, this hypothesis has never been verified through observation or experimentation. Furthermore, it is now well understood that structures which may be similar in appearance can be determined by totally distinct genes in different species. Thus, the genetic program being fundamentally different, it is a virtual certainty that the fundamental processes which follow from that genetic program, such as the stages of embryological development, will be very dissimilar, too. It has been proven that the embryological processes which produce similar-looking organs display many distinct aspects in each living being.

There are also huge molecular differences between living beings which appear to be related or analogous. For this reason, it is not even possible to talk about "molecular homology." Michael Denton's findings about this support what has been presented previously about molecular biology:

> Molecular biology has shown that even the simplest of all living systems on Earth have exceedingly complex structures quite unique to them…. In terms of their basic biochemical design, therefore no living system can be thought of as being "primitive" or "ancestral" with respect to any other system, nor is there the slightest empirical hint of an evolutionary sequence among all the incredibly diverse cells on Earth. For those who

hoped that molecular biology might bridge the gulf between chemistry and biochemistry, the revelation was profoundly disappointing.... Should this evidence in molecular biology have been discovered a century ago, organic evolutionary thought might have never been accepted at all. When there is no resemblance in molecular structures, embryological processes are different from each other, but different layers of structures can be substituted in the structures of similar organs.[129]

An important example is the astonishing resemblance of the eyes of various living beings and the observed parallels between the eye structures of very different animals. As a case in point, cephalopod vertebrates, like octopi and squid, and vertebrate animals and human beings, have no evolutionary connection between them—in other words, they are extremely different living beings. Further, there is no candidate with a similar eye to that of the human being and octopus which evolutionists could propose as a common ancestor between them because these two types of living beings are so far removed from one another biologically. Thus situated on the "evolutionary tree," these animals are said by advocates of evolutionary theory to have organs which are not "homologous" (similar and coming from a common ancestor), but rather, "analogous" (similar despite there being no evolutionary proximity). In other words, according to the supporters of evolution, the human eye and the octopus eye are analogous organs. Nevertheless, the organs that they simply consider as "analogous" are each resoundingly perfect, unique structures of such complexity. Although they resemble one another considerably in terms of their "camera technique," their retinas are very distinct. While the photoreceptor layer faces "the dark room," so to speak, on the octopus eye, it faces a totally opposite direction in the mammalian eye. Thus, it is completely unreasonable to claim that the similarity in these "camera techniques" of the octopus and mammalian eye occurred by random mutation. If the eye of the octopus truly had arisen by chance, as evolutionists say, then the vertebrate eye should have emerged via the exact same genetic incidents—the very same mutations, in other words. On the

other hand, one must also remember that in order for the position-
ing of the retinas to be unique to each species, as they are, the
occurrence of distinct mutations would have been required.

The evolutionist Frank Salisbury admits that even thinking
about the question proves to be a major headache: "Even something as complex as the eye has appeared several times, for example,
in the squid, the vertebrates, and the arthropods. It is bad enough
accounting for the origin of such things once, but the thought of
producing them several times according to the modern synthetic
theory makes my head swim."[130] According to the evolutionist
point of view, however, totally independent, random mutations are
supposed to strike identically and repeatedly at different times in
various living groups.

Another interesting example is the similarity between placental
mammals and marsupial mammals, so that marsupial ("pouched")
wolves, mice, squirrels, and moles all have placental counterparts
which exhibit similar morphologies. Evolutionary biologists believe
that two species in particular, namely the North American wolf and
the Tasmanian wolf, have completely separate evolutionary histories.
This belief is based on the fact that since the continent of Australia
and the islands around it split off from Gondwanaland (the supercon-
tinent that is supposed to have become Africa, Antarctica, Australia,
and South America), the link between placental and marsupial mammals is considered to have been broken—and prior to that time, there
were no wolves. However, the interesting thing is that the skeletal
structure of the Tasmanian wolf is nearly identical to that of the North
American wolf. Most notably, their skulls are witness to such an
extraordinary degree of resemblance that even specialists can barely
distinguish between the two creatures. Nevertheless, they belong to
different organizational groups entirely, as the former belongs to the
marsupial class and the latter to the placental class.

Accounting for the remarkable similarity of the Tasmanian and
North American wolf gives evolutionists problems, as the points of
resemblance between the two species have to be explained as being

a function of their derivation from a common ancestor, according to their "thesis." However, the truth is that marsupial and placental wolves are limited to entirely different continents and completely dissimilar environments. For this reason, these mammals, which have such parallel skeletal structures, should be claimed by evolutionist to have evolved separately via distinct processes—but then this itself would contradict their other claim that these similarities must have been transferred from a common ancestor through heredity. The end result of such convoluted evolutionary thinking has been a newly manufactured story whereby placental wolves and marsupial wolves are claimed to have been exposed to "similar evolutionary forces" due to "similar environmental conditions," whereby they independently developed "similar structures convergent with each other." Thus, in view of these types of "pairs" between placental and marsupial animals, wherein "parallel" animals have nearly the same morphologies, we can conclude that advocates of evolutionary thought uphold a model of so-called "convergent evolution," which claims the following: "The exact same mutations completely independent of each other must have produced these creatures "by chance" twice in different continents! Even if they were in different continents, they were developed by similar mutations which occurred at exactly same place, just like two people in separate continents, being quite unaware of each other, throwing a pair of dice millions of times in such a way that they both get exactly the same numbers, in exactly the same sequences."

Another important obstacle in the path of evolutionary theory is that both flying vertebrates and flying invertebrates possess wings. Actually if we ignore the feathers and finger bones of birds when considering that the bat's wing and the bird's wing are homologous, it can be admitted that there is a partial anatomical and embryological resemblance between them. However, the wings of flying insects and birds are completely different beyond the shared attribute of flight. Therefore, evolutionists call these wings "analogous," rather than "homologous," since they cannot make connections between them. How is it, then, that these strikingly similar structures which we call "wings," and which are used with remarkable effectiveness by creatures as varied as invertebrate flies and vertebrate birds (and whose principles are applied to human flight) could have emerged first? Consider that flies have no inner skeleton, but vertebrate wings have an inner skeleton. In both cases, nonetheless, the main goal is to succeed in the act of flight. The Creator, Who gave the lifting force to air, to permit flying in the first place, also gave wings to permit whatever creatures He so willed to fly out of their bodies. Just like we need knowledge and the study of aircraft engineering to build an aircraft, we need the One Who has control over both the air—to let birds, bats and flies fill the skies—and the embryological layers of each creature, with His Infinite Power and Knowledge. Otherwise, we would have to accept a truly irrational statistical event, one as unlikely to occur as the example mentioned above—that of millions of dice being thrown in different continents and yielding the very same numbers every time. Furthermore, the chance of such statistical concordance becomes even more reduced if one also considers the different types of flight. For every type of flying creature—flies, flying reptiles, flying frogs, flying fishes, flying mammals, birds, and others—has a particular mode of movement in the air, so that the dice in the example would have to deliver completely implausible alignments of numbers in succession in order to model the joint action of both natural selection and random mutations in the "evolution" of such variations in the act of flight.

6

From the Beginning of the Universe to the Chosen Earth

FROM THE BEGINNING OF THE
UNIVERSE TO THE CHOSEN EARTH

T he basic reason why evolutionary theory is based on chance, nature and causes—and why it is sometimes rendered as a worldview—is that it arises from both materialistic and atheistic philosophies. The idea that the emergence of living beings occurred only by means of evolution is suited to materialistic philosophy. Yet, if life "evolved" on Earth, the living conditions on this Earth needed to be favorable to allow living beings to survive on it. In that case, there must again be a Creator Who has infinite power and knowledge, so that an ecosystem which is ideally equipped with the necessary resources—like the air, water, sun and moon—could provide the best living environment for all kinds of organisms to exist on Earth in harmony.

The idea that the creation of life is solely based on material components and occurred by itself, as those components gathered together by chance, necessitates a huge assumption which forces us to include not only the world of living beings, but also the entire universe in our considerations. That is because for even the simplest-looking organic molecules to be synthesized, the necessary living conditions have to be prepared. Yet the issue is not just about simple organic molecules; rather, it is about complex living beings which exhibit the manifestations of infinite knowledge and power in each and every molecule, and which are perfect in all aspects. For such living beings to survive, very particular circumstances wherein all the necessary conditions are very precisely determined have to be pre-established. Therefore, in order for Earth to have developed in such as way that all the conditions are ideal for allowing living beings to inhabit it—and in order for the universe to have developed as it has, with such diversity and specificity, every moment since the Big Bang—then either nature (mindless, unconscious, and with unknown limitations), or an Infinite Power (Who rules over everything at every moment, at every point in the universe) has to render the service.

Despite the fact that atheists and materialists do not accept belief in God, they are nonetheless aware of the fact that they have to start by explaining the operation of evolution from the first moment of the universe. That is, they have to explain how the highly ordered universe evolved from a system where only random, chaotic, astrophysical processes were functioning prior to the organic evolution process from which they think living beings emerged.

There is a great relationship between the idea of evolution and the model of the universe as observed in the field of astrophysics. Critically, accepting that the universe has a beginning means simultaneously admitting that it was created—and a created universe is foreordained to be destroyed. Yet materialists, who essentially believe that the universe is eternal and everlasting, do not believe in either creation or destruction as necessities of their arguments. Further, according to atheists, if the universe does not have beginning or end, all kinds of emergences, originations, developments, and changes are to be automatically associated with the purported powers that are thought to reside in the universe itself, including natural forces and causes. According to their scenario, then, there is no need for a Creator with infinite knowledge and power. For this reason, real atheists do not accept the beginning or end of the material universe.

However, opposing this, advances in the fields of astrophysics, theoretical physics and quantum physics all point to the presence of a "creation moment for the universe." The Big Bang theory as it is defined today, the half-life processes of radioactive materials, and findings of cosmic background radiation collectively indicate a creation process for the universe and matter. Thus, materialists are forced to limit their focus to organic evolution, concentrating on how lifeless matter came to life—that is, advancing ideas about the process whereby creatures supposedly arose from dead matter.

Advances in astrophysics in the twentieth century brought forth two basic models of the universe. One of them was the "static universe" model and the other was the "expansion of the universe" model. According to the static universe model, there was no beginning to the universe; that is, creation could not be discussed as applying to the universe, and the universe was presumed to be eternal and everlasting. Needless to say, this idea was compatible with the fundamental beliefs which both materialists and atheists espoused.

On the other hand, the beginnings of the Big Bang Theory can be traced to the 1920s. In order for many aspects of the universe which Newton's "static and infinite" model could not explain to be elucidated through Einstein's "Theory of Relativity," the "Expanding Universe" model was developed separately by Georges Lemaitre and Aleksander Friedmann.[131, 132] Then, after Edwin Hubble's discovery that the light from stars was shifting to red—meaning that all the stars, with their galaxies, were getting farther away from each other—the expanding universe model was deemed to be both more plausible and reliable.[133] On the other hand, the materialist, Sir Arthur Eddington (1882–1944), rejected the Big Bang Theory totally because of his ideological point of view and atheistic beliefs, rather than because of any scientific opposition, by saying, "I find the idea that the universe has a beginning disgusting philosophically..."[134] In fact, the Big Bang Theory explained where the hydrogen that was required for the formation of stars (and which was not produced inside stars) came from. From this point of view, it addressed the criticisms of Hoyle, who had opposed the Big Bang Theory when it was first proposed, based on the problem of the formation of elements. According to the view of atomic particles prevailing in contemporary theoretical physics, very high temperatures were required to produce hydrogen. The Big Bang Theory accepts the existence of particular conditions—namely, exceedingly high temperatures and density—at the beginning of the universe. Fred Hoyle (1915–2001), however, proposed finding an alternative to the Big Bang Theory, as the Big Bang Theory necessitated the existence of a Creator. Hoyle, who was not willing to admit that life could not occur by chance, spent many years opposing the Big Bang Theory by saying things like, "Should the universe have begun with a hot Big Bang, then this explosion must have a remnant. Why don't you find a fossil of this Big Bang?"[135] Later, in 1964, upon the discovery by two radioastronomers, Arno Penzias and Robert Wilson, of weak electromagnetic radiation (cosmic microwave background radiation) coming from everywhere in space, the static universe model fell completely out of favor.[136] For this background radiation which was observed in the band of radio waves within the electromagnetic spectrum was nothing other than the waves that George Gamow had predicted based on the Big Bang Theory in 1948.

In 1964, Penzias and Wilson were working on the antenna of Bell Phone Laboratory to ensure communication with a satellite. While they were trying to measure the strength of the radio waves radiated by the Milky Way Galaxy at high-galaxy latitudes (beyond the plane of the galaxy), they discovered the temperature of the radio waves received to be a value equivalent to 3 °K (- 270 °C). Winners of the Nobel Prize for Physics in 1978 for this study, Penzias and Wilson had essentially made a discovery which was as important as that of the redshift in the electromagnetic spectrum was for astronomy (the notion of the expansion of the universe). Meanwhile, Big Bang theoreticians rediscovered the calculations of Gamow and his colleagues, who had predicted the existence of such background radiation as necessary remains from the initial creation moments of the universe, and whose temperature, caused by the expansion, was predicted in the late 1940s to be equivalent to 3 °K.

The existence of radiation of a very high temperature and very short wavelength in the first moments of the universe was essential in allowing Big Bang theoreticians to explain the abundance of hydrogen. This is because, having sufficient energy, this radiation would cause an increase in the amount of hydrogen by virtue of decomposing the heavier nuclei formed over time; then, while it would continue to be present following the initial expansion of the universe, the temperature would keep decreasing inversely, proportional to the size of the universe over time.

In short, the discovery of Penzias and Wilson was crucial, since it verified a phenomenon that had been predicted theoretically. Certainly, this discovery is the strongest finding which supports the Big Bang Theory, and it also shows that the universe, which initially had most of its energy in the form of radiation, acquired matter, in that most energy is found in the mass of nuclear particles.

At this point, it might be beneficial to discuss the Big Bang Theory in more detail—as it can even explain current ratios of chemical elements in the universe—in order to understand the phases by which the solar system, and our world, was created.

Current observations suggest that the creation of the universe began with a cosmic explosion which occurred at "time zero," some 15 BYA. This "Big Bang" is the creation moment of all measurable things—like time, space and matter. Under these supernatural conditions, four fundamental forces—namely, gravity, electromagnetism, and weak and strong nuclear forces—were possibly combined, and their strengths were the same. (Note that being one of the four fundamental forces, so-called "weak nuclear forces" are now understood to be part of electromagnetic forces as a result of the studies of both the Pakistani physicist, Abdus Salam, who won the Nobel Prize in Physics in 1979, and the Italian physicist, Carlo Rubbia, who won the Nobel Prize in Physics in 1984. Recently, "strong nuclear forces" have also been included with electromagnetic forces. Today, it is generally accepted that electromagnetic forces and gravity represent the only two fundamental forces. Thus, the long-time dream of physicists to express the active forces present in the universe by reducing them to only one—to create a "grand unified theory"—might finally be coming to an end.).

According to the most common models of the universe, the universe was marked homogeneously and isotopically with an extremely high energy density, and incredibly high temperatures and pressures. Then, approximately 10^{-35} seconds into the expansion, the universe began cooling rapidly (its temperature decreasing billions of degrees in a period of time as brief as one billionth of a second)—and it was then subject to a sudden expansion during which it grew exponentially. It has been predicted that the universe grew in volume at an incredible factor of 10^{150} within a very short period of time, which is estimated to have happened between 10^{-35} second and 10^{-32} seconds into the expansion phase. Nonetheless, its size was not bigger than an apple yet. This phenomenon, named "inflation" by the astrophysicist, Alan Guth, is understood more clearly when compared to a kind of phase transition where all proportions are conserved, for example, when a drop of water suddenly evaporates and fills a larger space.

The universe possibly reached an average rhythm of expansion while passing from 10^{-32} second to 10^{-12} second. Temperatures were so high that the random motion of particles occurred at relativistic speeds, and particle-antiparticle pairs of all kinds were continuously created and then broken to pieces through collisions, to form light particles and photons within their little world. Then, the photons quickly became particles and antiparticles again. According to the calculations, first the "strong nuclear forces" were separated from the other fundamental forces during the expansion; then, the electromagnetic force and the "weak nuclear forces" were separated at the 10^{-12} second mark. So then, there were four types of forces which came into play, as the universe entered a new energy phase. At this point, fundamental particles known as "quarks" started moving in this "energy sea," and electrons, neutrons and their antiparticles converted into matter (i.e., transitioned from energy to matter).

The size of the universe was possibly about the size of the solar system at 10^{-6} second after the initial expansion, and it continued cooling until its temperature reached a couple of billion degrees. Particles became more mixed and tended to be more stable. Under such conditions, quarks were able to gather together, antiquarks were able to regroup, and new types of particles, like baryons and antibaryons, were formed by the action of the strong nuclear forces. Further, the small excess of quarks over antiquarks led to a small excess of baryons over antibaryons.

Approximately 10^{-4} seconds (ten thousandths of a second) after the Big Bang, the universe was probably filled with photons and light particles, or light antiparticles—that is, electrons and positrons (anti particles of the electron), and neutrinos and antineutrinos—as the temperature was no longer high enough to create new proton-anti-proton pairs (similarly for neutron-antineutron pairs). As a result, a mass annihilation immediately followed, leaving just one in 10^{10} of the original protons and neutrons, and none of their antiparticles. A similar process happened at about 1 second, this time affecting electrons and positrons. After these annihilations, the remaining protons, neutrons and electrons no longer moved relativistically, and the energy density of the universe came to be dominated by photons (with a minor contribution from neutrinos).

Presumably, at the moment when the universe was one second old, its temperature was 10 billion degrees and it no longer contained antimatter. Matter was composed of protons and electrons, which balanced each other—that is, the universe was electrically neutral—and neutrons were ten times fewer in number. Everything else was light. There were approximately one billion photons present for every particle of matter. Frequent collisions between the fundamental particles occurred; repeatedly a proton and a neutron combined to form a deuteron (a heavy hydrogen nucleus), the simplest of nuclear systems. In turn, the deuteron was sometimes broken up by a photon. The universe, being full of radiation, was opaque in appearance (resembling a dark, adhesive fluid), as photons were confined by matter particles. Then, one second after the birth of the universe, the flow patterns of events started to change. The temperature dropped to a billion degrees; the heat energy fell below the level of weak forces; and deuterons were no longer able to break up, so their numbers started to increase. Then, those deuterons combined with protons and neutrons to form helium nuclei. This was the first nucleosynthesis, and it took a couple of seconds. The universe, at that moment, is thought to have been composed of helium nuclei (^{4}He) and some other light nuclei (deuterium, ^{2}H; helium, ^{3}He; and lithium, ^{7}Li).

After that, the first "expansion crisis" started. The temperature dropped 100 million times below the temperature of the center of the Sun. The huge energy which resulted ensured the creation of more particles and antiparticles, following each other in close succession, over very short intervals of time. By expanding about 1,000 times more, the universe began to fill a space as big as the size of today's solar system. Free quarks were held within the neutrons and protons. Then, after this thousand-fold expansion, protons and neutrons combined to form atomic nuclei, which comprise the largest portion of today's helium and deuterium. All these events are estimated to have happened in the first minute of the expansion. Meanwhile, as the universe filled with energy, it warmed temporarily, and this caused the expansion to stop.

Nevertheless, since the temperature was still very high, the conditions required for atomic nuclei to capture electrons were not present yet. After the expansion continued for about 300,000 years, neutral atoms—which balanced the positive protons by capturing electrons—emerged in wide distribution; yet the size of the universe at that point was still considered to have been about 1,000 times smaller than it is today. Later on, neutral atoms started to gather inside gas clouds in order to form stars. The universe expanded up to one-fifth of its current size; stars became clustered as groups, which could be deemed young galaxies.

Then, when the universe reached about half of its present size, a significant portion of heavy elements, which typically form planets like Earth by means of nuclear reactions within stars, were produced. If prevalent calculations are true, the Sun took shape six billion years ago and the solar system formed five billion years ago, when the universe reached two thirds of its current size. Indeed, the number that can be given through these findings regarding the Earth's age is closely related to the model of Earth's creation. Should the Earth have been created as a result of the accumulation over time of some larger pieces which were formed earlier—a process called "accretion"—both the age of each piece, and the timing of their combining to compose the present globe, could also be discussed. However, it is not easy to determine whether or not the fusions that occurred during the accumulation (accretion) process completely wiped out any footprints related to the age of origination of the first pieces. Even if there are places where the footprints were not erased, and even if these can be discovered (they are expected in the crust—or rather, inside the earth), then the age of the samples taken from those areas might possibly point not to their initial time of creation, but rather to the time of accretion. Therefore, the large number obtained may not actually indicate the age when the world took its ultimate shape; this age might finally turn out to be even bigger.

In the meantime, the creation of stars within a certain period of time caused the gas reserves in galaxies to become exhausted. So, the number of newly formed stars began decreasing. Then, within the next two or three minutes, the temperature decreased to a billion degrees. The fusion tendencies between protons and neutrons, under the influence of strong nuclear forces, began to occur. The first atom nuclei created in this way had very short lives; a proton and a neutron were promptly combined in those nuclei, resulting in formation of a deuteron, which was then easily scattered by photons.

In order for the temperature to drop from a couple of billion degrees to a couple of thousand degrees—and for the heat energy to get close to that required for the action of the electromagnetic force—a million-year period was probably required. (Note that the electromagnetic force is a million times weaker than weak nuclear forces) At this stage, a hydrogen atom could be created through the fusion of a proton and an electron; and in the meantime, a photon, which could break up a newborn atom, would be emitted. The temperature continued to drop.

When the temperature dropped below ten thousand degrees (300,000 years are assumed to have passed by then since the Big Bang), photons could no longer obstruct the formation of atoms. Under the influence of the electromagnetic force, each isolated proton (a positive charge) captures an electron (a negative charge) to make a hydrogen atom by bringing electrons and protons near to each other; and each helium nucleus (two positively charged protons and two neutrons) connects to two electrons to form a helium atom.

Atoms are transparent to most photons. Thus, photons gain freedom of movement by separating from the matter which confines them. The universe is suddenly lit with a sparkle of light. Radiation, sending beams of the same intensity in all directions, overruns the universe.[137]

At about 3,000 degrees, it was likely that each proton was stably surrounded by an electron, and each helium nucleus was surrounded by two electrons. A new stage, the birthing time of atoms had arrived. However, the bond between the proton and the electron in the hydrogen atom was not saturated yet; two hydrogen atoms could combine to form one hydrogen molecule. In other words, the birth of the atom and that of the molecule occurred almost simultaneously. Before the birth of atoms, space was full of electrons, and this seriously obstructed the scattering of light. But upon the capture of electrons by atomic nuclei, the universe became transparent, and light started traveling across the entire universe, unobstructed by any obstacles. Fossil radiation, at 2.7 °K, quite possibly started at this moment. The development of the universe then calmed down over the next couple of million years due to the stability of hydrogen and helium as compared to nucleosynthesis.

Photons lose their energy more and more over time, and deuterons live long enough to capture an extra neutron and a proton; as a result, fully stable helium nuclei are created. These are then charged with functioning as the basis for new atomic formations. In the universe whose volume is continually increasing, matter is spread in a way that does not provide the opportunity for particles to gather and combine. There is only a faint electron and photon mist where helium nuclei and free protons (they are potential hydrogen atom nuclei) float. This indicates a universe containing perhaps a dozen hydrogen nuclei to every one helium nucleus; that is, one fourth of its mass will consist of helium and three fourths will consist of hydrogen. The observations of astrophysicists verify these theoretical findings which support the "Big Bang" notion.

This process presumably continued for tens of thousands of years (with respect to our time measure). Photons, which have the energy to obstruct the fusion of atomic nuclei with the electrons in the universe, continued to expand and cool down. However, they were always confined inside matter—they could not separate from the mass of particles and become free. Thus, darkness and opacity still prevailed in the universe, as though it were covered with a dark veil.

The stage where stars were formed is reached by the development of gravitational force, the final fundamental force. In other words, as soon as the universe had attained a stage wherein matter was ready for a new level of organization, there appeared a new manifestation, the only physical force which was capable of being appointed as such a servant, to control everything in the range of causes.

Unlimited amount of material started to be combined to form the first galaxies, and the homogeneous universe started to become heterogeneous. Possibly differing in mass, the first stars, termed the "first generation," took shape as a result of the influence of the forces of attraction on the primary material (hydrogen, helium, and lithium) within the constitution of those galaxies. Being a hundred times more massive and a hundred thousand times brighter than the Sun, some particular stars became extinguished three or four million years later; these are called "blue giants." The others were smaller and had the capacity to live for billions of years. Then, over time, the consumption of interstellar material gradually reduced the possibility of the formation of new stars.

A second chance was created for a new nuclear stage in stars where the material became shrunken and warmed due to the strong attraction force which was particular to them. The temperature increased at the center of the star to reach ten million degrees, and collisions became so strong that hydrogen nuclei underwent fusion to form helium nuclei. Here, the stability of helium was seen once again. But the material was not scattered; instead, it became dense, and thus, the Sun was created as a result of a nuclear fusion reaction. At this stage, the system calmed down; helium was produced by using hydrogen, and its geometry was fixed (to the same radius and the same brightness). Actually, it almost represented the very condition of the Sun today. For the Sun has been shrinking for fifteen million years (the "Kelvin-Helmholtz" or "T Tauri" stage), and if current age determinations are correct, hydrogen has continuously been transformed into helium for 4.6 billion years.

If doomsday does not occur before then, the Sun is predicted to entirely deplete its reserves in five billion years under normal conditions, and its center will be transformed into helium. As a result, it will return to its initial shrunken form. Being older than the Sun, and having experienced these stages before, stars serve as "factories" for producing the heavy elements required for the formation of a planet like the Earth. As a result of this shrinking process in the stars billions of years ago, a temperature increase commenced due to the effect of attraction, and it reached a hundred million degrees. This high temperature caused three helium nuclei to undergo fusion to form a carbon nucleus (^{12}C), and here began the second big stage in the creation of the universe. Gradually, the shrinking slowed, the atmosphere of the star expanded, and the star transformed into a giant red mass (a red giant).

To summarize briefly, matter was organized following the Big Bang, and it was differentiated through various stages. The "red giant" phase, where the temperature was in the order of 100 million degrees, and which followed the "Sun" phase, was the most productive and richest stage in terms of producing almost all of the chemical elements which exist within the lifetime of stars. Small molecules dependent on carbon were created in the interstellar medium that absorbed the material which supernovas emitted. Therefore, we see that stars are born, grow and die—just like other living things; that is, they, too, undergo an energy crisis.

Hydrogen and helium were created in the severe temperatures of the Big Bang 15 billion years ago (if age determinations are right). Being the vital elements of living organisms, the more complex atoms, such as carbon, oxygen, calcium and iron, were created in the very hot depths of stars as a result of nuclear processes—that is, in the most unfavorable conditions for life.

The elements which were created by massive explosions were ejected into interstellar space later and transformed into new stars and planets as a result of the attraction force in that medium, as electromagnetism was converted into the chemical ingredients of life. In short, everything from the ink on this page to the air we breathe was created from this first generation of stars, in the range of causes.

Explosion: Bull's Eye

What would happen if a bomb were to explode in any locality, or in the building where we live or work? Everything would break into pieces; all objects would fall into disorder and become disarranged, and doors and windows would be damaged, and so on. In short, destruction and disorder would result from the huge amount of energy released by the explosion. In fact, all types of uncontrolled energy discharges cause catastrophes and disorder similar to what results from the detonation of bombs.

If one were to insist that an explosion on a scale like the Big Bang, which engendered the orderly creation of the entire universe, occurred without control, supervision, or consciousness, and without an Organizer, one would have to accept incredible and unreasonable impossibilities. There is no room for chance or coincidence when one executes the countless probabilistic calculations which would be required to explain the wide range of critical effects: for example, the cooling of temperatures from trillions of degrees to a measure of warmth which allows us to live; the creation of fundamental particles, atoms and molecules; or the accommodation of each and every star, among millions of stars, each in orbit with a certain balance of attraction, within each and every galaxy, among billions of galaxies.

The particular arrangements which arose 4.6 billion years ago and which allowed our lives—namely, the forces of attraction; the critical rotational velocities of planets in particular orbits, one of those being our own planet, Earth; and the intensification of gas and dust clouds required for the creation of the Sun, the most important star for us among the 10^{11} stars present in one specific region of the Milky Way Galaxy—all demonstrate infinite power and foresight, and are the most obvious and miraculous evidence of the art of One God, Who is the owner of infinite power, knowledge and wisdom.

It is irrational and foolish to claim that so many critical conditions—such as the Earth's being furnished with the most ideal attributes so as to be favorable for life; the specific distances of the Moon and the Sun to the Earth; the unique composition of the climate, soil, air, and water; and the particular presence and action of carbon atoms, as well as elements such as nitrogen, hydrogen, and oxygen which bond with carbon to become the foundation of organic life and, hence, the basis of all living organisms—were somehow assembled by chance and accidentally resulted in the birth of the universe. Accepting that this universe and everything in it emerged from the random consequences and brutal force of a colossal explosion, and interpreting all scientific groundwork in the name of evolution—simply for the sake of not associating this whole perfect system with a Creator whose power and knowledge is infinite—is an absurd stance which an intact mind, reason, conscience and heart could never approve nor accept.

THE CHOSEN PLANET

According to our knowledge today, we do not think that there is biological life in a form we know of on any planet or star other than the Earth. However, it is surely possible for God to create specially equipped creatures suitable to the characteristic conditions of other planets or stars, with His infinite power and knowledge. The preparation of innumerable factors—such as the particular concordance of Earth's distance to the Sun; its rotational velocity; its orbit; the placement of the Moon as a satellite; the specific amounts and densities of gases in the atmosphere; the distribution of heat, climate, rainfall, wind, and mountains; and so on—to achieve the most suitable and unique substrate for the continuation of life is a manifestation of the immeasurable knowledge and power that embraces the entire universe.

The presence of the atmosphere is a crucial precondition for the created biosphere, while its destruction would bring the extinction of living organisms. The current state of the atmosphere depends on two factors: the optimal distance between the Earth and the Sun, and the chemical density balance between carbon dioxide (CO_2), oxygen (O_2), and ozone (O_3).

A regulating, causal role was given to the atmosphere itself during the creation process. If the amount of carbondioxide increases, then the temperature on Earth also increases (the "greenhouse effect"). In addition, by absorbing the excess amounts of carbondioxide, oceans become more acidic ($CO_2 + H_2O <=> HCO_3^- + H^+$); to overcome this effect, the rates of decomposition of rocks (chemical change) and vegetation growth both increase. In this way, the atmosphere has continued to function as a system in a state of theoretical equilibrium for at least 300 million years (note that there is no practical equilibrium simply because of relatively minor "ebb and flow" fluctuations around the theoretical equilibrium).

Consequently, the atmosphere is the finest, most delicate and most important attribute which plays a role in the development and perpetuation of the biosphere. Yet its establishment was almost "instantaneous" in terms of the geological time scale, as it was made out of gases—essentially, out of continuous activity and changes over very short intervals of time. While it was of an acidic and reductive character initially, it became an oxidizer as a result of the establishment of photosynthesis. This oxidizing quality of the atmosphere initiated an external geochemical effect which encircled the Earth. Being present as a barrier, or shield, between the Earth and the vastly different conditions which exist beyond it (consider, for example, that there is a density of 10^3 molecules/cm^3 in space, compared to a density of 10^{18} molecules/cm^3 on Earth), our atmosphere has a critical causal role in protecting us from the lethal radiations of the Sun. At this point, the importance of the arrangement of the "atmosphere-hydrosphere-geosphere-biosphere" system for life to have emerged and to continue with such perfection in the long run—as compared to the disorder which prevailed at the beginning—becomes extremely clear.

7

From an Inorganic to an Organic World:
Dice Turning Up Double Six Each Time

FROM AN INORGANIC TO AN
ORGANIC WORLD: DICE TURNING UP
DOUBLE SIX EACH TIME

T he reason why evolutionary theory and creation touch on
the same points from time to time as a necessity of the
field, and why they reach different interpretations of the
same evidence, is based on how two vital concepts—"intention"
and "perspective"—shape worldviews and belief systems. A person
who starts the journey of discovery with the intention or preconception that the universe does not have a creator or owner, and
looks at each event with that worldview, can produce very different
scenarios according to his or her own belief in the intention and
appearance of things, by referring selectively to all kinds of observations and information.

Furthermore, even though the initial conditions on Earth—and
all sorts of claims about the origin of life under such conditions—
necessitate proof, no interpretation can actually go beyond being a
scenario which is unproven scientifically. Given this, and as a fundamental requirement of objectivity and ethics in science, all types
of claims should be examined, no matter how far these may seem
to be from scientific criteria. Some of the conditions or factors
which are mentioned in those scenarios are even likely to have
played a role in the chain of causes as "veils" over the divine names
of God during the miracle of creation—in other words, as veils of
material causation which originate in the will of the Creator. The

existence of a chain of such material causes and effects neither keeps creation from being a miracle, nor obstructs it. Rather, such a chain of astounding events in the material world makes it even easier for us to grasp how miraculous creation really is. We are living in the world of causes, and our Creator may have used all these material causes (elements, heat, light, charges), which belong to this world, to cloak His glory and greatness. Yet, since the manifestation of His power and knowledge are essential in the act of creation, the idea of material causes being a "veil" should not be emphasized. The real miracle is that those causes were predestined and chosen expressly to yield favorable conditions, in appropriate amounts, and at the right time, to become a "life soup"—and later on, all living beings were created.

Thus, rather than debating or denying the possibility that partial truth might be present in the scenarios "montaged" to account for the intention and appearance of the first atmosphere, first oceans, and first land conditions, such truth can simply be used to prove the existence of the Creator, Who has infinite power and knowledge—rather than to deny the Creator.

The most important indicator that creation is a miracle is that both appraisal (of requirements) and selectiveness (of best conditions) are exhibited in the ordering of all causes in particular amounts, and in their arrangement in specific, consecutive sequences. On the other hand, some might argue that the material infrastructure of the act of creation might have been prepared by simply applying or combining all of the environmental conditions and factors (all the causes together), and thus have somehow "arisen" in the first days of our planet. However, what we should keep in mind here is that merely the special preparation of this material infrastructure over a certain period of time is not sufficient for the act of creation to occur. Life does not emerge without a special "life impetus" outside of matter itself—that is, life does not occur without the manifestation of the Name of God, "The Life-Giver."

After all these considerations, we may take a look at some of the information which has been discussed in various fields of science relating to the possible causes for the emergence of the universe—even though we see these as shadows compared to the real truth—in order not to be the object of blame, nor to be accused of being unscientific or of being an enemy of science.

Our planet is estimated to have had no free oxygen initially, and so, no protective ozone layer in the upper atmosphere. The energy that was necessary for the biological synthesis of this layer, as an ordinary cause in the process of creation, might have come from the ultraviolet radiation of the young Sun, or from electrical discharges in the first atmosphere—and it might have come from an unknown source that we cannot imagine. Nonetheless, the important thing is that the presence of this energy was in "the ideal dose," and of "usable quality." Since this energy was without conscience or reasoning, and the boundaries of its strength were uncertain, it would not have been useful for anything other than destruction and eradication—in other words, it is impossible for the energy required for biosynthesis to emerge autonomously, or by chance—without conscious knowledge. Even today, the mysteries of the structures of cell organelles—such as the mitochondria, which are vital for the energy metabolism of living beings—is not properly understood. One must be ignorant of probabilistic calculations to think that respiratory enzymes and coenzymes, mitochondrial DNA and other enzymes required for biosynthesis, somehow evolved by themselves.

Having infinite power, it is possible that our Lord could have prepared the environment and conditions for living beings before He created them. According to our time measurements, we estimate that the improvement of the Earth's conditions prior to the creation of living beings probably took approximately four billion years. In the meantime, He might well have started life in the oceans simply to provide the first organisms with protection from lethal ultraviolet radiation. The creation of life on land, on Earth, perhaps occurred at the end of the Devonian period (about 400–530 MYA), since this arrival of land animals may also have coincided with the beginning of the ozone layer.

From Inanimate to Animate

The biggest problem for evolutionary theory is the origin of life. The difficulty of explaining how a creature that could be called the "first animate being" emerged from a mixture of inanimate elements remains an obstacle which is impossible to overcome. All of the claims about how the transition from the inanimate (inorganic) world to the animate (organic, sentient, growing, and behaving) world occurred cannot go beyond being merely hypothetical.

The cell, being the fundamental structural unit of all living beings, is a complex machine made up of about one trillion atoms. How the transition from atoms and molecules to the first cells took place is still not known at all. Furthermore, we do not actually know whether "a gradual transition from inanimate to animate" life happened in the first place. The claim of such a "gradual transition," of molecules being organized into a living cell, developing step by step, seems to be a scenario montaged for the benefit of the evolutionary hypothesis. Since we cannot credit millions of molecules with gathering and thinking together, and deciding collectively to get organized as a cell, and since atoms lack mind, consciousness, reasoning and knowledge, then if we do not accept the existence of a Creator, we are forced to admit that all of the amazing elements and functions of the cell are simply caused by chemical reactions—of unlimited power, and in unrestricted amounts—which somehow yield convenient, random results.

In fact, not much of the information obtained from research performed on fossils of bacteria sheds light on the origin of life. It had been generally understood for some time that they had cell walls covering them as the presence of a cell membrane is the only solution which can account for the protection of the cell's internal regulatory processes and the passage of matter in and out without upsetting essential systems which are intrinsic to conditions on Earth. Since the fundamental structural units of living beings are amino acids, which compose huge organic molecules—proteins— then the initial conditions on Earth and the atmosphere must have been suitable for forming these molecules. Unfortunately, the biggest mistake made here is supposing that by knowing what kind of material is used in some artwork, we can instantly know how such material was processed, and how the artwork was made. The same mistakes are made about the first creation. Discovering the organelles in living cells, knowing some of the macromolecules which are placed inside their structures, apprehending some of their chemical properties, and discovering the specific elements and the amounts of those elements in their structures—even all these together do not indicate in the least how cells became animate in the first place, that is, what kind of creation processes they underwent.

In 1932, J.B.S. Haldane (1892–1964) and the Russian biologist, A.I. Oparin (1894–1980), attempted to perform experiments to determine whether or not carbon-based organic compounds in the first atmosphere—which they accepted as having no oxygen—could be produced. Oparin argued that, as simple inorganic compounds mixed together over time, they formed more complex organic compounds. Then, over longer periods of time, they formed the first living organisms, which were claimed to be heterotrophs, that fed on the organic compounds deposited in oceans; thus, according to this framework, the first plants did not use photosynthesis to produce their food. However, questions about how the first cellular-type systems formed and reproduced, and how the complex proteins and enzymes on which they depended arose, remained unanswered. The idea that clays—attractive three-dimensional structures—might have played a role as "models" or "molds" in the first development and polymerization stages of organic molecules could not pass beyond being merely a claim, remaining completely elusive. As the originator of those ideas, Oparin stated that lipid polymers (fat molecules) had the ability to be folded and formed into hollow spheres (coacervates), and thus they might have formed an "inner medium" which provided an opportunity for the first metabolism—yet he could not show how those fat molecules arose on their own and formed a membrane with proteins.

According to Oparin's argument, amino acids were combined in a certain order and system, depending on differences in their shapes and electrical charge distributions, and thus they formed complex molecules. Those molecules later caused the formation of "buds" on microscopic water drops, all on their own. To verify his claim, he did experiments on microscopic units composed of glue and gelatin, following his own line of thinking, from the first cellular model that he accepted, one which was presumed to be of a gelatinous composition.[138] Yet even though he added enzymes externally (of course), he was not able to obtain anything to verify his idea that it was possible to step across the huge gap between being inanimate and animate, to "create" a living being; that is, he could not show how, or for what type of purpose, mindless and unconscious molecules could have gotten together to create perfect, vital, complex structures.

Inspired by this idea, chemist Harold Urey (1893–1981) thought that the first earthly atmosphere could have been similar to Jupiter's atmosphere, which is composed of a mixture of ammonia, methane and hydrogen. In 1952, by adding water, which is vital for life, to a set-up that had the ingredients of the first atmospheric conditions, Harold Urey and his student from Chicago University, Stanley L. Miller (1930–2007), attempted to see if organic molecules could be formed by chance occurrence. The two scientists stimulated a chemical medium through a glass loop containing ammonia, methane, hydrogen and water vapor, using electrical sparks, ultraviolet radiation and electric current (to simulate lightning in the atmosphere), in order to observe whether or not amino acids would be produced.[139, 140] Twenty-four hours later, they discovered that along with many other compounds, glycine, aspartic acid, glutamic acid and the alanine amino acids were formed. The synthesis of those organic molecules was announced to the world as if life had been created from nonlife, giving the clear impression that all questions were answered and all problems were solved, as the purported "solution to the chemical evolution problem" was served up to the public. Jeremy Rifkin, epigrammatizes this as follows:

> With great fanfare, the world was informed that scientists had finally succeeded in forming life from nonlife, the dream of magicians, sorcerers, and alchemists from time immemorial. Since that historic occasion, virtually every biology student has been made privy to the wondrous secret Miller and Urey had uncovered, a secret that had eluded humanity's grasp over the ages. Great comfort is taken in knowing finally where life originated. In fact, so intent was the need to resolve this question of origins that little effort was extended to probe some of the basic assumptions underlying the Miller/Urey experiment. Had the scientific fraternity bothered to exhibit even a bit of healthy skepticism, they would have seen, at the time, that the Miller/Urey experiment was as much a fictional account of genesis as the long-held myth of spontaneous generation by which scientists of an earlier age had claimed that life arose from dead matter by observing maggots mysteriously appear out of garbage.[141]

In fact, Miller's experiment contains many critical inconsistencies. First, he thought that he was imitating primordial Earth-like conditions, but he used a mechanism called a "cold trap" for the experiment done in Urey's laboratory. Without a doubt, such a cooling, protective isolation mechanism did not exist in the primitive Earth's atmosphere. Second, Miller preserved the amino acids by isolating them from the environment as soon as they were formed. Since there was no such isolation mechanism in the primordial Earth's atmosphere, very severe and difficult conditions in the environment in which the amino acids were formed would immediately have destroyed these emerging molecules. In sum, the cold trap mechanism requires intelligent design, and it is not reasonable to presume the presence of such isolating, cooling, and protective functions, intended to shield nascent forms, in a primitive environment where ultraviolet radiation, lightning, high oxygen ratios and various toxic chemicals are considered to have existed. The chemist Richard Bliss expresses this contradiction by observing, "Being the crucial part of Miller's set up, the cold trap was used to isolate the products formed by chemical reactions. Actually, without this trap, the chemical products would have been destroyed by the energy source."[142]

Another weakness of this experiment was the neglect of the fact that hydrocyanic acid, formic acid and, especially, nitric acid would quickly and easily be formed in the same medium. Further, when accounting for the sulphuric acid formed by hydrogen sulfite—which was mixed into the atmosphere by volcanic explosions that were thought to function as storage units for solar energy by emitting ultraviolet radiation reaching 240 nanometers in wavelength—in addition to the presence of all of these acids, each with a destructive and disruptive nature, what would have been formed was nothing but a burning mixture which would not have been favorable to life at all.

Besides, another real-life danger for the amino acids which were artificially obtained in the experiment was hydrolysis. Indeed, amino acids which are simply put into a test tube with similar conditions simply disintegrate into smaller molecules, such as cyclic anhydride, glutamate, aspartate and pyrrolidone, in water.

Debates about the early nature of the atmosphere have brought the subject matter of chemical evolution to a deadlock. The ideas about this are a serious subject of discussion between evolutionist biochemists and geologists. The presence of limestone ($CaCO_3$), which was deposited billions of years ago, is held by geologists as evidence for ammonium not being present in the same medium, since the pH values of ammonium and $CaCO_3$ compromise each other. Had methane actually existed in the early Earth's atmosphere in great amounts, we should have determined this through geological observations. In addition, if such an atmosphere had existed earlier, hydrophobic organic molecules, protected by sedimentary clay layers, should have been found. Yet while abnormal amounts of carbon and organic molecules have been identified in old rocks, such hydrophobic organic molecules have never been noted. Further, the early atmosphere hypothesis, which presumes a composition of methane and ammonia gases, can also be understood as baseless and unsound from the fact that methane and ammonia have not been observed to come out of volcanoes.

Biochemist Peter Mora, of the National Cancer Institute, in the US, says the following in regard to the experiment: "There is a great deal of controversy on this score: in fact, so much controversy that in the final analysis, any experiments designed to duplicate the primeval environment are no more than exercises in organic chemistry."[143] Therefore, even though Miller and Urey's exercises in organic chemistry seemed convincing in the beginning, after careful analysis, they turned out to be of absolutely no scientific value in terms of addressing the question of the origin of life.

Yet speculations pertaining to Miller and Urey's experiment are numerous, indeed. The Belgian biochemist, Marcel Florkin, says, "The idea of a primitive reductive atmosphere has been abandoned"; and "It is considered to be insufficient in terms of geological evidence."[144] In any event, geochemists have now come to an agreement that Miller's experiment pertaining to the early atmosphere of the Earth was not prepared realistically. Furthermore, many scientists think that the primitive Earth's atmosphere consisted of volcanic gas explosions which included water vapor, carbon dioxide, nitrogen and little bits of hydrogen.[145, 146] Indeed, pioneers of origin of life studies, Sidney Fox and Klaus Dose, agree that Miller, "used the wrong mixture of gases in his experiment." Scientists also agree that the free hydrogen in the early atmosphere would have diffused easily out of the atmosphere, and the remaining methane and ammonia would have been oxidized.[147] In his recent study, Holland explains that there are two basic opinions concerning the composition of the early atmosphere. According to the first one, which he agrees with, there was either no, or only very little, oxygen in the early atmosphere. Conversely, according to the second one, which the majority of the scientists agree with, there was a great deal of oxygen.[148]

Therefore, the studies which consider the Miller and Urey experiment to be particularly invalid from the start are those which relate to the presence of oxygen in the early atmosphere and photolytic reactions. Evolutionist biochemists accept as preconceived dogma that the early atmosphere did not contain oxygen, for if there had been oxygen, oxidation would have occurred and amino acid synthesis would thereby have been obstructed. Nonetheless, according to a significant percentage of geologists, the early atmosphere did consist of a high amount of oxygen (at least 200 billion tons). Brinkman, a geologist, argues that there was so much oxygen in the early atmosphere that it would not have permitted biochemical evolution to happen.[149] It is also possible that the Earth's atmosphere might not have changed much over time, since rock formations contain oxidized iron. That indicates the presence of an oxygenic atmosphere for the primitive Earth. Also evidence of "the Earth having an oxygenic atmosphere since the time of the oldest rocks, 3.7 billion years ago" has been found.[150]

While Miller and Urey's idea of an oxygen-free (reductive) atmosphere was struggling to overcome this important barrier, it ran across a second barrier which was impossible for it to pass. For if there had been no oxygen, there would have been no ozone (O_3) layer either; thus, the amino acids would have immediately been destroyed, since they would have been exposed to the most intense ultraviolet rays, without the protection of the ozone layer. These rays, coming from the sun or other sources, cause chemical decomposition (photolysis and photodissociation). Therefore, life could not have emerged, even in the most primitive form, under such earthly conditions—that is, in the absence of oxygen.

a)	H_2O ------- ultraviolet -------> radiations	$OH\bullet + H\bullet$
b)	$H\bullet + H\bullet$ --------------->	H_2
c)	$OH\bullet + OH\bullet$ ------------>	H_2O+O (atomic oxygen)
d)	$O + O$ ----------------->	O_2 (molecular oxygen)

The possible process of the formation of oxygen in Earth's early atmosphere

In short, both the existence and nonexistence of oxygen is a handicap for evolutionists. R. L. Wysong explains their quandary as follows: "If oxygen were in the primitive atmosphere, life could not have arisen because the chemical precursors would have been destroyed through oxidation; if oxygen were not in the primitive atmosphere, then neither would have been ozone, and if ozone were not present to shield the chemical precursors of life from ultraviolet light, life could not have arisen."[151]

In order to eliminate this problem, the idea that life initially developed under water—so that it was thus protected from the killing ultraviolet rays hitting the Earth—was proposed. But, at that point, a third barrier (which was actually much larger than the first two) arose, since there would be no possible energy catalyzer. This is vital for the Miller and Urey experiment, as they used electrical discharges to activate chemicals and argued that lightning would have done the same job in the real world—yet lightning would not have been able to penetrate the water which covered the Earth, and their experiment included both ammonia and methane. For even in the event that lightning were able to penetrate through the water (which it actually cannot), the chance of any biological formation occurring automatically as a result of this would have been zero. That is because, in order for life to have begun in this way, water vapor, ammonia, carbon dioxide, nitrogen, and methane would have had to produce amino acids under water, and then these amino acids would have had to combine to form polypeptides, also under water—but that is where the problem becomes completely unresolvable since polypeptides cannot undergo synthesis when there is an excess of water in their environment.

Ammonia is also very sensitive to photolysis and decomposes into its components, nitrogen and hydrogen, when it is subjected to ultraviolet radiation. Water molecules decompose into hydroxyl and oxygen under the influence of ultraviolet rays, too. The released oxygen molecules combine with methane to produce carbon dioxide and water, and they also combine with ammonia to yield nitrogen and water. As a result, the initial structure of the early atmosphere would have been transformed into a mixture of CO_2, hydrogen, nitrogen, and water vapor.

The existence of ammonia in the primitive atmosphere is evidently very crucial in order for organic molecules to have formed, as nothing could be obtained from the experiments done later without the use of ammonia. Yet even if there had been ammonia gas, many studies of its decomposition due to ultraviolet radiation— given that it is very sensitive to photolysis—have revealed that all of the ammonia would have decomposed into hydrogen and nitrogen within 30,000 years, according to Abelson, or perhaps as long as 500,000 years, according to Ferris and Nicodem.[152, 153]

When the non-presence of ammonia in the early atmosphere came to be strongly supported and generally accepted, experiments started to be done without using ammonia. However, the results were consistently negative: neither amino acids, nor even their subgroup molecules, aldehydes, could be obtained from those experiments. Then, in 1975, two American scientists, Ferris and Chen, repeated Miller's experiment many times using an atmospheric environment that contained only carbon dioxide, hydrogen, nitrogen gas, and water vapor, and they were unable to obtain even a single amino acid molecule.[154] They were only able to get some alcohols, acetone, ethanol and formaldehyde. Yet in the end, Miller's experiment continued to gain a lot of attention through the efforts of certain groups, while Ferris and Chen's findings were hardly mentioned at all. Interestingly, in 1985, at the "Molecular Evolution of Life" symposium, Miller himself confessed that his experiments concerning the early atmosphere could not be accepted as being realistic due to the fact that ammonia would have dissolved in the oceans, so that a surplus of ammonia could not have been present in the early atmosphere.[155] He also said that there was no scientific reason for choosing methane and ammonia gases—that it had just been his personal preference because he would not have been able to obtain any amino acids without ammonia. In addition, another of his confessions can be mentioned here: "There is not any consensus concerning the composition of the early atmosphere. Since rocks older than 3.8 billion years old are not known, there is not any evidence about the conditions on Earth between 4.6 and 3.8 billion years ago."[156]

Even though A. Katchalsky's experiment with aminoacyl adenylates was reported to have been successful in producing sixty or more units of polypeptides using nickel and zinc, along with montmorillonite, a common clay mineral,[157] those polypeptides could not possibly have escaped destruction in the early conditions on Earth and in the atmosphere—that is, under the lethal effects of ultraviolet rays with wavelengths of 250–300 nm and 300,000 joules of energy. Moreover, it is also inevitable that nickel and zinc would form other compounds with nitrogen, nitric acid, and chloric acid in the extremely high temperatures of primitive Earth. We should also keep in mind that the chance of the existence of the exact and particular lab conditions which were artificially applied to the actual, early atmosphere—as well as the presence of elements like nickel and zinc in the precise, specified amounts which were used—is nil. Thus, critical questions are left unanswered, like why the reaction would ever be started with such a molecule as aminoacyl adenylate, in an environment where there was neither technology nor a natural system; how the obtained matter would ever have been protected from thermal entropy in reality; and how the regular energy necessary for the formation of high-energy bonds between the atoms of organic molecules could have occurred during a time which predated photosynthetic reactions.

In conclusion, McMullen's statements about the weaknesses of the Miller and Urey experiment should sound very reasonable to most people:

> The last and most formidable weakness of the Miller experiment is Miller himself. He designed the experiment, hoping to produce amino acids, but the first run did not generate any. It was back to the drawing board. He changed certain experimental parameters and the second run did provide the desired results. Now a supposed strength of the experiment is that it is a possible naturalistic explanation of the origin of life. The methane, ammonia, water, and hydrogen in the Miller experiment, even though of an artificially high purity, could be the Earth's early atmosphere. The electric spark could be analogous to lightning, and the liquid water, the oceans. If so, then what is the analogy for Miller, the designer and modifier of the experiment? The answer is an intelligence—a designer; God, if you will, is needed for life to occur. If one thought the earlier inferences from the Miller experiment were scientific, then one has to concede that this inference of a supernatural being is also scientific.[158]

In order to prepare a background in which life can be generated, particular forms of amino acids first have to be obtained. Amino acids are divided into two groups: levorotary (left-handed) and dextrorotary (right-handed). These two amino acids are complete mirror images of each other, much as one's left and right hands are the same but opposite, and this feature is called chirality. Dextrorotary forms are incapable of supporting life—in fact, they are often lethal. Thus, the amino acids of all living forms are levorotary, that is, left-handed. Right-handed molecular forms are only found in DNA and RNA; all the other components of living beings are built from left-handed amino acids, other than a couple of exceptions, such as the exoskeleton of insects. Applying this information, Wilder-Smith points to another failure of the Miller and Urey experiment, as the special amino acids which Miller and Urey claimed to produce in their experiment were not suitable for the formation of life. Wilder-Smith makes the points as follows: "For biogenesis to take place, all building blocks (amino acids) of living protoplasm must be levorotary.... If even very small amounts of amino acid molecules of the dextrorotary type are present, the proteins of a different three-dimensional structure are formed, which are unsuitable for life's metabolism."[159]

The deadlock which materialists face here is that in all of the experiments in which they expected to produce the "life soup," as it were, they obtained 50% levorotary and 50% dextrorotary acids, which then formed molecules called "racemates," or a racemic mixture. In fact, racemates are incapable of synthesizing life, but the Miller and Urey experiments produced only racemates. In fact, every experiment of a similar kind has produced only racemates; and as Wilder-Smith points out, a racemate is not, under any circumstances whatsoever, capable of forming living proteins or life-supporting protoplasm of any type. We should emphasize that until this time, it has proven absolutely impossible to form anything other than racemates by stimulating nonliving chemicals with electrical discharges. Harold Urey was asked at a conference, "If you could, explain how life could have been formed by the chance combination of chemicals, when all living things require pure levorotary amino acids, whereas in laboratory experiments such as yours, only racemates are produced by spontaneous processes?" His reply is worth repeating: "Well, I have worried about that a great deal and it is a very important question ... and I don't know the answer to it."

Some other experiments have been done which argue that amino acids, formed by chance, came to be deposited; and then the favorable ones gathered in orderly sequences, and this process produced proteins. Those experiments have a significant role relating to organic evolution. As the primary building blocks of all living beings, proteins have been targeted by evolutionists, who seek to prove that they would have formed—and did—by chance, on their own. But this has also turned out to be an essential problem for advocates of evolutionary theory to overcome.

The process of bonding hundreds and thousands of monomers, as a result of one molecule of water being released from a carboxyl group of amino acids, to allow amino groups to form longer chains of peptide bonds, is called "polymerization." In turn, proteins are complex molecules which are built up through the bonding of hundreds or thousands (depending on their sizes) of amino acid molecules as a chain (polymerization). In general, a chain which is made of about 100 amino acids is called a "polypeptide," while polypeptides which are composed of more than 100 amino acids are called "proteins." Further, in order for a molecule consisting of a huge chain to be considered to be a protein, it has to play a role in the living cell, taking an active part in certain structures, like enzymes, hormones, or nucleoproteins. In this respect, proteins are also fundamental molecules of the cell—building blocks of living mechanisms. There are proteins as big as one thousand, ten thousand, and even a hundred thousand molecules.

The most widely known experiment concerning proteins was that conducted by Sidney Fox (1912–1998). He wanted to check if proteins could have been formed near volcanoes in early Earth conditions. By heating dry amino acids in a test tube for 4–6 hours at 150–160 °C, he obtained a simple molecule "pile," similar to a protein, that he called a "proteinoid." On the other hand, the fundamental neglect of some critical points—like his use of pure, dry amino acids (they would have broken down had they been wet), which actually could not possibly have been deposited on the primitive Earth; and his exposing them to heat only for a very short time (they would have been burned and spoiled on the early Earth due to exposure to extreme temperatures over a long period of time) caused Fox's experiment to lose its strength. Another weakness of his experimental procedure was that these molecules, which he called proteinoids, were like random spots, and quite dissimilar to the proteins of living organisms; in any event, it would have been impossible for them to be protected from disintegration in the Earth's early conditions. Most importantly, they were deprived of any genetic system that could reproduce them. Yet, against their opposition—who argued that ultraviolet rays would decompose such newly-formed proteins—supporters of organic evolution claimed that those proteins were formed under water, which therefore allowed them to be protected. But in that case, Fox's experiment becomes completely meaningless since he expressly used only dry amino acids. Besides, any reaction which releases water (amino acids release water when forming proteins) does not seem likely to occur in water, according to "Le Chatelier's principle." A water-releasing reaction does not occur in a medium where there is already water and the reaction itself is a reversible process. Therefore, rather than forming a protein out of amino acids in an aqueous medium, the reverse effect occurs; in other words, if a protein is put into an aqueous environment, it will break into amino acids. In closing this point, we can simply conclude that water obstructs the formation of proteins.

Even though he is an evolutionist, G. A. Kerkut sums up the state of science when it comes to speculation over the formation of the first living being:

> There is, however, little evidence in favor of biogenesis and as yet we have no indication that it can be performed. It is therefore a matter of faith on the part of the biologist that biogenesis did occur and he can choose whatever method of biogenesis happens to suit him personally; the evidence for what did happen is not available.[160]

In conclusion, Miller and Urey's much-fussed-over experiments are of absolutely no scientific value in addressing the question of the origin of life. Similar to so many other speculative attempts that have characterized the evolutionary literature, their study—if it proves anything at all—shows how hard it is to support a theory that is confused at each step of the way by a reality that firmly refuses to be adapted to its governing hypothesis.

Another unfortunate misconception is to expect that a polypeptide chain formed by chance might trigger the origin of life.

Polypeptides are precursory molecules which have not become proteins yet. Proteins, which are large organic compounds, are made of polypeptides, which fold at certain points and get a particular, thickened shape. Being composed of about twenty standard amino acid molecules, proteins play key roles in countless processes which allow and affect life in all living beings. A protein structure has four distinct aspects, namely its primary structure, secondary structure, tertiary structure, and quaternary structure. Certain numbers of amino acids are found in each protein molecule, and these are arranged in a sequence which is unique to that protein. This amino acid sequence is the primary structure of the protein, and it defines both the shape and function of the protein. The angles between the peptide bonds that connect the amino acids in the molecule chain determine the secondary structure; hydrogen bonds generally cause the molecule to take on a spiral shape. The tertiary structure is formed by the twisting and folding of the protein chain; it is generally stabilized by nonlocal interactions. In some proteins which are formed of more than one polypeptide chain, like hemoglobin, the forces of the ionic bonds deriving from electric charges, which are characteristic of the tertiary structure, determine the arrangement of the polypeptide chains, or the quaternary structure.

It is worth expanding on this point in detail: one can imagine taking a long chain in hand and first folding it into two, then twisting it, and then twisting it again from another region until it became untwistable. The result in one's hand would be a particularly-shaped iron form. Just like the twist of this iron chain, then, proteins also twist at certain regions and fold on top of each twisting, resulting in some very uniquely shaped structures, such as hemoglobin.

Proteins can be classified into two groups as *"proteins which take part in structures,"* and *"proteins which play a role in biological or physiological activities."* However, some proteins are both structural and functional. Most structural proteins are composed of long fibrous chains. For example, collagen, which is found in bones, tendons, cartilage and connective tissues, and keratins, which are found in various parts of the body, like the skin, hair and nails, are structural proteins. Conversely, those proteins which function in biological or physiological activities are mostly spherical in shape; these include, for example, diverse enzymes which catalyze chemical reactions; hormones, which serve as messengers between various parts of living mechanisms; carrier proteins; and antibodies.

Even the most minor error during any of the numerous folds mentioned above renders a nonfunctioning protein molecule. The position and order of amino acid chains in all proteins is determined by the sequence of DNA nucleotides. When the synthesis of a certain protein is required, the unique code, which is present in the DNA for that protein, is transmitted to the nucleotides on the RNA molecule. Each of the three nucleotide groups determines a distinctive amino acid; and in any case where the string of amino acids, which is sequenced with respect to the order in the RNA code, gets out of order somehow, various disorders and defects appear. There are, on average, between 400 and 3,000 amino acids in protein molecules, and the molecular weight of proteins generally varies from 100,000 Da (one "dalton" equals 1g/mol) to 500,000 Da, and may even reach one million Da.

A change in the location of a single amino acid in the polypeptide chain, or an absence or excess of only one amino acid in the chain, makes the entire chain nonfunctional, resulting in many diseases and bodily malfunctions. For instance, in the hemoglobin A molecule, which is made up of 574 amino acids, and has a molecular weight of 68,000 g/mol, the replacement of only one amino acid—valine with glutamine—changes the fundamental characteristic of the entire molecule and causes a very serious disease called sickle-cell anemia. Or, there might be a mistake in the synthesis of the enzyme responsible for folding, due to a rupture in the DNA or a missing gene in the DNA. However, when living beings do not exist yet, and when DNA and RNA are not present yet, protein folding to obtain a particular conformation is not possible, as the very enzymes which take part in folding—and the DNA encoding them—do not exist yet. If that were the case, both the proteins and the enzymes—as well as the DNA and RNA molecules themselves—would have to be assumed to have formed simultaneously, strictly by chance. But no mathematician would ever accept such a probability.

The problem of how to close the gap between inorganic, non-living things and the first "living" creature is the most difficult problem both for evolutionary theory and for the philosophy of biology. In spite of the innumerable probability calculations proving that proteins and nucleic acids cannot simply come into being by chance, some will never give up and they make statements like "Even if the possibility seems to be zero as to probability calculations, that does not mean that it is impossible..."

Now, for an instant, let us just assume that a protein has come into being by chance, and then let us see if the first living being could have arisen by chance from this, or not. First of all, in order for this being to be called the "first living being," it would have to possess at least some of the basic characteristics of life. Such a creature, even it were only single-celled, would require a sufficiently complex system displaying the fundamental characteristics which distinguish living beings from nonliving, such as alimentation, growth, a specific shape and size, internal organization, being open to stimulation, and engaging in metabolic activities, including reproduction. Carrying out many functions as essential requirements of being alive necessitates special structures in the cell called "organelles," which each represent the finest artwork. Each organelle is designed to execute a particular duty: for example, mitochondria are centers for energy production; golgi apparatuses produce necessary secretions; ribosomes synthesize proteins; lysosomes perform cellular digestion; centrosomes and microtubules carry out cellular division; chloroplasts are the center for food production in plant cells; chromatin carries nucleic acids where the genetic code, which has the position of central control mechanism in each cell, is encoded and packed. The most important characteristic of all of these structures is not only that they incorporate, or use, many enzymes in the activities they perform, but also that they can produce those very enzymes. Furthermore, each type of organelle is built in a very precise and unique form.

Enzymes are biocatalytic molecules which enable biochemical activities to be carried out more quickly, efficiently, and smoothly, and in ideal conditions. For example, a chemical reaction which occurs at 700–800 °C in a laboratory environment can occur at 37 °C in the presence of a catalytic enzyme. As a case in point, carbonic anhydrase—an enzyme which decomposes carbonic acid during respiration into water and carbon dioxide—can break down, or decompose, 500,000 molecules in one second. In turn, proteins constitute the foundation of enzymes, which then render a service in all sorts of biological activities, from digestion to respiration, and from circulation to sensory processing. Moreover, certain coenzymes (unique molecule groups which are in key positions) work with some enzymes in order to allow them to function optimally. The structure of coenzymes generally consists of some vitamin derivatives and nucleotide units.

Since enzymes are basically made of proteins and synthesized from proteins, a program or a code is vital for their synthesis, and that code exists in nucleic acids, both DNA and RNA, which are in the position of controlling the cell. Except for in some viruses, a DNA molecule functions as a "chief control center," so to speak, and an RNA molecule functions as an "execution center"—where the translation of the instructions is achieved, so that synthesis is performed with respect to the commands given by the DNA. However, this fact raises an important problem, which is that both DNA and RNA also need enzymes for their own synthesis and reproduction. Thus, we are faced with two processes which necessitate each other: nucleic acids are needed for the synthesis of enzymes, but enzymes are needed for the synthesis of nucleic acids. Now, we not only have the problem of trying to account for the synthesis of enzymes, by chance, or for the further synthesis of complex organic molecules, by chance, or for the further preparation of the entire program of a living being from nucleic acids, also by chance—but we are critically confronted by a scenario which is impossible even to imagine, which is that of the simultaneous co-occurrence of two such utterly unlikely coincidences.

Just to avoid or eliminate this difficult problem, theoreticians of evolution, who had already become aware of the impossibility of the sudden emergence of a cell by chance, started to argue that being the frontier of the cell, coacervates and microspheres formed first, and then they somehow "transformed" into cells. According to them, proteinoids—assumed to have formed by chance—constituted a system over time by diffusing into a water drop whose exterior wall somehow started to function as a cell membrane. However, the selective permeation characteristics of the cell membrane, its extremely perfect structure, and the behavior of many special receptor molecules within it, is still not fully understood even today; the special structure of transit regions is employed as a very sensitive doorkeeper, and the three-layered membrane model proves that the cell membrane is a microcosmos in itself. Those who claim that glicolipids and special integral proteins in the fluid mosaic membrane model—which is itself made of special protein molecules placed between two-layered phospholipid molecules— formed by chance simply testify to their own ignorance of molecular biology.

Yet according to the advocates of evolutionary theory, enzymes, which are themselves supposed to have formed by chance, somehow passed through this excellent membrane, which is also supposed to have formed by chance, and thus located themselves inside of those drops. Then, as soon as the DNA chain, which is also supposed to have arisen by chance, started to function inside that water drop, a living creature emerged. Even though critical questions remain unanswered—By which kind of mechanism did this water drop, this coacervate, start to reproduce? How were its energy needs supplied? How was its DNA and RNA coded? and so on.—

such a miraculous living being can seemingly still arise "by chance" according to evolutionary thought.

The fact is that in spite of the presence of the very significant technological advances which are available today, the possibility of ensuring so many varied experimental conditions in the laboratory, and the supply of all sorts of organic molecules, even from other living beings, scientists have failed to make a cell, in all of its aspects.

Some evolutionists argue that RNA molecules, in the fashion of "naked genes," might have been the first precursors of life. Since the DNA molecule chain consists of two strands and has a more unique structure than RNA, it becomes more feasible to start with the idea that the RNA molecule chain, which contains only one strand, formed on its own. On the other hand, questions about how the first RNA molecule would ever have started "making its own copy," and how programs and enzymes for complex activities—such as reproduction, metabolism and growth—were formed simultaneously, by chance, again fail to be answered.

Another typical "way out" which serves the evolutionists' pre-judgment about the origin of life is viruses. Since they do not have a metabolism nor the characteristic of being stimulated on their own, viruses may seem to be "nonliving." Upon entering a living cell, they function and reproduce as parasites, using the enzymes of the host cell. Thus, in order for viruses to function as living beings, they need a fully functional living cell which they can enter. So, we are right back where we started. Furthermore, consider that these "simple-looking" organisms, these viruses, also have a genetic system which is composed of nucleic acids and proteins; and consider, too, the weakness of human beings against the many diseases they cause. Reflecting in this way, it becomes possible to understand that viruses also have a very complex structure, even as single-celled living organisms—one that cannot have formed by chance.

Concluding on this point, even though it has been repeatedly stated that various organic molecules, such as viruses, proteins and nucleic acids, cannot be formed on their own by chance, the claim that "coincidence" and "chance" could form a living being has always been brought to the table. However, there is another way of proving that any kind of useful organic molecule cannot be formed on its own by performing probabilistic calculations in which all sorts of circumstances are accounted for. So, let us go ahead and explain the impossibility of forming a living being by chance with answers to the following questions.

Each living cell has the logic of an amazing program—a feature which is referred to as "irreducible complexity." Each of the organelles of the cell is made up of particular molecules, in very precise amounts, having a perfect arrangement and function. Let us try to understand the trouble which evolutionary theory has with irreducible complexity by supposing, for an instant, that those molecules were formed by chance. If we look at the advances made in biochemistry, microbiology and cytology over the past fifty years, the articles and books written relating to the cell would be too numerous to fit into most modern libraries. Every day, our knowledge about the cell deepens and intensifies, and we are faced with increasingly interesting results; however, when we turn back and look behind, to see how much progress we have made, we sometimes get the feeling that we have "barely moved an inch." This is the same effect, in fact, that we experience when we feel as small as a pebble as we get close to a mountain that once looked so small from a distance—for the deeper we go into the intricate functioning of cell, the dizzier we become. We are astonished before the infinite knowledge and power that manifests itself in this magnificent artwork, which demonstrates both a conscious plan and exemplary outcomes.

Volumes of books could be written only to show how the idea of cellular evolution contradicts basic reason and intelligence. When we consider the biochemical processes of a human organism, in which trillions of cells serve critical objectives in a consistently harmonious way, we begin to witness and appreciate the spectacular systems inside the cell.

Focusing specifically on only one point, the well-known American biochemist, Michael J. Behe, in his book, *Darwin's Black Box*—which relates the impossibility of evolution with respect to biochemistry and microbiology—clearly shows the molecular and chemical dead-ends of evolutionists in detail. As a matter of fact, every scientist who believes in the Creator could write many books simply based on the valuable information he or she can gain from that book, in which are featured many typical examples of irreducible complexity in common biochemical and microbiological events which occur in our bodies and surroundings every single day. For instance, consider the human eye; against the proven "irreducible complexity" of the eye, in terms of its molecular and biochemical processes—which are "cloaks" of causes which describe the miracle of sight through anatomical, histologic, physiological and embryological data—not a single advocate of evolution can offer a reasonable or convincing explanation.

Claiming that amazing biochemical and microbiological processes might have evolved gradually, in order—or that the "package" of such metabolic processes could have evolved in big jumps somehow, based on what was needed—may seem like a pleasant idea, but it is supported by neither the molecular structure of life nor the principles of biology. While clearly explaining that delicate parts, such as the fine structure of the flagellum or cilium, the perfect tail-like "motor" organelle which ensures that a unicellular organism can be in motion—which includes considerations of the complex fibers in the structure of these tail-like organelles, the annuli, the hook-like projections, the mechanism for converting a sliding motion to a bending motion, and sophisticated microtubules—could be designed and composed only by virtue of infinite knowledge, Behe states that ciliates and flagellates, which carry more than 200 proteins just on their tail-like organelles, obviously disprove evolutionary theory. Such a molecular machine simply does not work unless all of the constituent parts are present; that is, ciliary motion does not occur if the microtubules, connectors and motors do not exist. Thus, we can cite the example of the tail-like organelle of flagellum or cilium as something which is too complex to have evolved from simpler predecessors and is at the same time too complex to have arisen through chance mutations.[161]

Another piece of evidence which Behe presents to explain and substantiate the idea of irreducible complexity is the vital phenomenon of blood clotting, which can only be considered the artwork of a Power with consciousness and infinite knowledge, and which exhibits the importance of turning to biochemistry and molecular biology for a correct understanding of the degree of refinement in evidence in the type of processes which the blood goes through before clotting, and the specific enzymes and factors which are secreted at each phase, at precisely the right time, in exact amounts, by particular cells—as if the cells and organelles themselves were aware of how to behave when bleeding occurs.[162]

Behe explains many other wonderful phenomena relating to the cell in a very striking fashion, such as the movement of matter in and out of the cell through particular channels in the cell membrane; the functions of each organelle in the cytoplasm occurring as part of a marvelous program; the motions of microtubules and fibers; the immune system's development of immunity against microorganisms entering our bodies; and the impossibility of RNA and DNA forming by chance.

All this information obtained from the microscopic world, then, effectively voids the input of those who offer apparent and superficial similarities—purportedly attained as "evidence" of evolution from the fields of comparative anatomy and embryology.

8

Probability Calculations

PROBABILITY CALCULATIONS

T
he delicate regulation required for life to emerge and continue on Earth, the regulation which exists in the Milky
Way, and which includes not just the Earth but the Sun
and Moon, as well, has been the subject of much research. According to those studies, in order for any type of life to exist on a
planet, satellite, star or galaxy, the environment has to have certain
attributes which are determined within very narrow parameters.

Let us mention them briefly. First, an insufficiency or excess of
any attribute can cause many life-threatening problems. A few
examples point to the critical importance of factors such as the type
of galaxy; the relative distance of supernova explosions, and their
frequency of occurrence; the other planets which comprise the
remainder of the system; the distance or closeness of stars to the
center of the galaxy; the number of stars in the planetary system
which give birth to planets, and their relative age, size, color, and
brightness; the surface gravity; the inclination of the orbital plane,
the relative eccentricity of the orbit; the inclination of the rotational orbit, and the time needed for it to rotate on its own axis; the
age of the planet; the thickness of its crust; its magnetic field, the
rate of light reflected as a function of total light; the rate of incidence of meteorite and comet impacts; the ratio of oxygen and
nitrogen in the atmosphere; the levels of carbon dioxide and water
vapor; the ratio of electrical discharges; the level of ozone; the
amount of oxygen; seismic activity; the ratio of oceans to continents; the distribution of continents on the sphere; the specific

minerals in the soil; and the forces of mutual attraction between the moon and planet. All these conditions had to be set to the most ideal standard in order for the Earth to have become a suitable place for living mechanisms.

Since everyone accepts that celestial bodies lack the willpower, intelligence and consciousness to ensure this arrangement on their own, there is no other possibility than to believe that they either obtained their present position and composition by chance, or they were created by the will of a Creator with infinite knowledge and power. For this reason, evolutionists refer to the concepts of probability and coincidence, and they make these notions the basis of their worldviews.

The fact that astonishing numbers which the mind cannot even grasp are regularly presented by countless researchers in various subjects—using mathematical theories and calculations related to the study of probabilities—is ignored by the advocates of evolutionary theory; in this way, events which are claimed be the outcomes of probability and chance are often falsely portrayed as phenomena which can occur very easily. However, the fact is that the probability of the emergence of even the smallest attribute of any biological being based on chance has a probability of zero, and this is clearly seen if one just takes a brief look at a couple of examples of the sorts of probability calculations which have been performed by researchers, some of whom believe in evolution, and some not.

Here, it is worth mentioning some of the many probability studies performed by Hoyle, Crick, Guye, Morowitz, Salisbury and, most importantly, Coppedge, in order to examine the subject matter in detail. Emerson Thomas McMullen summarizes some of the calculations arrived at by those scientists as follows:

> I once entered the *Sports Illustrated* magazine sweepstakes. If I had won, they would have paid me one million dollars, tax-free, in twenty-five installments of $40,000. In the fine print, the magazine said the odds of winning that year were one in 1.2 x 10^8. This means, on the average, I would win once every 120

million years. Let's say I happen to live for the next 120 million years and the contest is conducted each year. Normally I would expect to win just once. What do you think the chances are for me to win the grand prize each and every year for the next 120 million years? Sounds impossible? According to Sir Fred Hoyle and others, I have a fantastically better chance of winning the *Sports Illustrated* Sweepstakes 120 million years in a row, than of life forming on earth by naturalistic means. Hoyle and Wickramasinghe calculate an extremely low probability for the formation of an enzyme: one in $10^{40,000}$ - that's 10 with 40,000 zeros behind it. Winning the *Sports Illustrated* contest 120 million years in a row has a probability of only 1.44 in 10^{16}.[163, 164]

Thus, even if the entire Earth were nothing but an "amino acid soup," the occurrence of such an event would be virtually impossible. Hoyle also gives the example of the Rubik's Cube: in order for this "toy" to properly align itself on its own (so that each face would be of only one color), even if it made a random move every second, would take 1.35 trillion years—meaning that using chance alone to execute this relatively simple task would require a duration which is 300 times longer than the actual age of the Earth.[165] So, the real question is no longer whether or not evolution is possible—but whether or not it is probable. Even when taking into account the fact that the universe is estimated to be 10 billion years old, Sir Fred Hoyle (1915–2001), in his book, *The Nature of the Universe*, declares that this still does not allow enough time for the chance evolution of the nucleic codes for each of the 2,000 genes that regulate the life processes of the more advanced mammals. He points out that believing that chance occurrences of random mutations, over a long period of time, accidentally created the complex and orderly relationships which are expressed in genetic codes is akin to believing in the probability that "a tornado sweeping through a junkyard might assemble a Boeing 747 from the materials therein." In fact, Hoyle believed in the idea that life came from space, from beyond Earth ("panspermia"), and that evolution was governed by "intelligent design"—and he vehemently opposed Darwinism and the idea of biochemical evolution on Earth. In turn,

Francis Crick (1916–2004), one of the discoverers of the double helical structure of DNA—a man who did not believe in Creation—also arrived at an extremely low probability for life to have originated naturally.[166]

Based on the oversimplification of two kinds of atoms which are ordered in proteins, Charles Eugène Guye (1866–1942), a well-known Swiss physicist, found a probability for their arrangement of 2.2×10^{-320}. He also established that the probability of obtaining a simple protein molecule from 40,000 atoms of five elements—like carbon (C), hydrogen (H), oxygen (O), nitrogen (N), and sulfur (S)—was 10^{160}.[167] This was reported by Pierre Lecompte du Noüy in *Human Destiny* (1947) as meaning that 10^{243} years would be required for one single protein molecule to be formed by chance.[168] Yet, since the longest proposed ages for the universe and the Earth are 10^{11} years and 5 billion years, respectively, and life requires more than one protein, what we face is nothing other than an impossibility.

The genetic programs of higher-order complex organisms contain information equivalent to a billion bits, or the letter sequences of a small library consisting of a thousand books. (Note that the information in the genome of higher-order complex organisms is still not fully known, though recent studies have shown that the human genome contains more than a billion bits of information; however, even if only one tenth of all DNA is messenger DNA, the problem remains.) These genetic programs contain commands which trigger the growth and development of billions of cells to form a complex organism, and they also contain thousands of algorithms, as coded forms, that specify and regulate specific commands with respect to particular tissues and organs. According to Denton, even for a skeptic, in terms of reason alone, it is shameful to believe that those programs came into existence only by means of a chance process.

The advocates of chance evolution apply simple probabilistic calculations, like tossing a coin or die, to the cell, organelles and other organic molecules, starting from the formation of the simplest protein—but they fall back on "time" when they are faced with completely improbable numbers at the molecular level. However, the calculations which account for the ages of both the Earth and the universe obviously preclude the use of "time" as a solution to the problem of "evolution by chance," and thus show the failure of arguments for evolution based on probabilistic calculations.

To demonstrate, let us consider the probability of the chance occurrence of a protein, enzyme molecule, organelle, or cell, which is small in the beginning. Further, let us suppose that it did happen once that a living cell emerged by chance, like a "one-time lottery." However, the subject matter is not limited to this, for advocates of evolutionary theory argue that it is necessary to base the evolution of all living beings—all of their tissues, organs, metabolic processes, anatomical systems, and the entire being, in fact, perfect in all aspects and started from a cell—upon the same concepts. Further, according to them, the role of coincidence and chance is not limited just to this—for all living ecosystems, every living-nonliving relationship, the entire Earth, the solar system, and the whole universe, are presumed to have formed through such sequential chains of chances. In short, they assume that all of life—from the human brain, and its consequent humanity and civilization, to the universe—everything, in fact, is the "art of chance." In such a world, where everything is founded on chances and coincidences, would there be any need for God, religion, the inner heart, or ethics?

In the event that the bottommost levels of hierarchical systems, from atoms to galaxies, depend on chance, there is no doubt that the uppermost levels would consequently be given up to nothingness and dereliction. Thus, since the building stone of living organisms is the cell, and the building block of the cell is protein molecules, it is important to emphasize the probabilistic calculations of whether or not a very simple protein molecule could ever really form by chance.

Should the possibility of coincidence and chance forming structures favorable to a particular purpose, according to a certain plan, be analyzed with respect to the simplest molecular level, then making a decision about the relative chance of the "upper levels" emerging, or not, would become easier. If we separate all the carbon, oxygen, hydrogen and nitrogen atoms on Earth into appropriate ratios in the most useful way we obtain 10^{41} groups. Admitting that 30 quadrillion reactions would occur in each group, and working with a rapidity rate that would form 10^{24} distinct chains in one year, then a total of 10^{65} chains would form in all amino acid groups in one year. Now, assuming that this process had been going on for 5 billion years, which is generally accepted as the age of Earth, this would mean that 10^{75} different chains would have formed since the origins of the Earth. This number, at first glance, might seem very big, so some might think that it would be possible for one protein to form within this probability range. However, if we dig into the structure of proteins in more detail, it will quickly be seen that the calculation is actually not that simple.

In order to determine how many different chains could form from 20 amino acids on a protein molecule, each consisting of 400 amino acids on average, we would need to calculate the 400th power of 20—that is, 10^{520}, with 520 zeroes following the number 10. That is to say, a very large number of possible combinations would need to be considered. Thus, the occurrence of only one useful protein arising from these randomly sequenced chains would be one in 10^{240} (which is also the probability of writing a meaningful word of 400 letters using a 20-character alphabet). Now, assuming that all the atoms on Earth make amino acids, the occurrence of 10^{75} distinct chains having arisen since the beginning of the Earth has already mentioned above. So, in order to find out how many useful protein molecules would form among these many chains, we need to divide the last two numbers, which gives us a result of 10^{-165}.

Thus, there is no further need to perform the probability calculations for a protein with 574 amino acids, as about 3 trillion hemoglobins—which blind chance can clearly not make—are formed in our bodies every second.

Dr. Harold J. Morowitz, of Yale University, calculated that in order for even the simplest living being to survive, it would need 239 different types of proteins. However, a living being so simple is not known to exist today. *Mycoplasma hominis* (H 39), known as one of the smallest bacteria, has 600 kinds of proteins. So, could the simplest living being, embodying such huge, complex molecules, truly be formed as a result of coincidence?

Earlier, the chance of even one useful protein being formed using all of the appropriate atoms on Earth was calculated to be one in 10^{165}. Similarly, when we think about the chance occurrence of 239 proteins forming separately, and then combining by chance to form a complete living being, the probability reaches incomprehensible levels. Without further drawing out this point, it can be stated that the probability of a complete living being forming by chance is the number arrived at by expressing one quadrillion to the power of 9,975, that is, $10^{119,701}$. Morowitz, in his book, *Energy Flow in Biology*, calculated the probability of chance fluctuations generating sufficient energy for the bond formation that molecules need in a living cell. Even with an ocean of the correct molecules, which are necessary to make the simplest cell, the chance of their bonding properly would be one in $10^{399,999,866}$.[169]

Let us continue thinking of much simpler cases. Let us imagine cutting 10 identical circles the size of a metal coin, on average, out of cardboard; writing numbers from 1 to 10 on each coin, and then putting all of them in a small bag. After mixing them properly, the probability of pulling out on the first trial the circle on which the number 1 was marked is 1/10, as all circles are identical and chosen randomly. If one puts each circle back into the bag after it is pulled out, the probability of drawing the numbers 1 and 2 successively is 1/100. Thus, if one intended to pull out all the numbers from 1 to 10 successively, and one were to assume that the simple process of pulling out each circle takes only 1 second, then in order to succeed with a 100% guarantee, this person would have to be ready to work on this activity for 317 years, day and night without stopping—

clearly an unreasonable timeframe for completion. That is because the probability of randomly pulling out the numbers from 1 to 10, one after another, is as small as 1 in 1×10^{10}. Yet if it is so difficult to obtain a sequence consisting of only ten elements, then forming protein chains composed of thousands of amino acids in the same manner, by chance, would surely be much more difficult— impossible, indeed.

Let's leave the probability of the formation of such a protein molecule by chance aside for a moment. If we examine the probability of typing a two-word phrase, "fossil records," consisting of 14 characters (13 letters and 1 space character), by chance, a very different picture will appear. The probability of randomly typing the phrase, "fossil records," using a 27-letter alphabet (26 letters and a space character), is about 1 in 109 trillion. Analogous to the calculation of a physics professor at Yale University, William R. Bennett, if a person were to type one random character per second using an alphabet of 27 characters, it would take approximately 48.5 billion years for him to type "fossil records" only once.

Now, let us go further and suppose that all of the carbon, nitrogen, oxygen, hydrogen and sulphur atoms—which are found on the Earth's crust, in water, in the air, and in the structure of amino acids—have already formed the amino acids completely. That is, taking the number of all of the atoms of these elements into account, 10^{41} possible amino acid units, each containing sufficient amounts of 20 different amino acid types, would be available for reactions to make protein. In living cells, if we accept the duration of protein synthesis in each unit to be 5 seconds, on average, then each unit could make 6,372,000, which would then yield 6.3×10^{47} amino acid chains in a year, from those 10^{41} units. Also, suppose that the entire Earth—essentially, a huge laboratory—started to function right after the world was created, and that it had been fully operational for 5 billion years. In this case, based on the calculations just mentioned, a total of 3.15×10^{57} amino acid polymers would have been synthesized as a result of 5 billion years of hard work.

Now, let us consider, too, that the two basic characteristics which determine the specific kinds of proteins which result from synthesis are the types of amino acids they contain, and the order of amino acids on the chain (even if they are of the same sort). Thus, if X_n represents each amino acid, then a protein consisting of the amino acid chain, X_1-X_2-X_3-X_4...X_{100}, is said to carry distinctive features compared to a protein which is made up of the chain, X_2-X_1-X_3-X_4....X_{100}, and so on. Earlier, when first defining the concept of a "protein," it was stated that a molecule consisting of at least 100 amino acids, which functions as a structural element, enzyme, hormone, or nucleoprotein, could be considered to be a protein; but amino acid chains which do not play a role in the structure of any cell or contribute to any kind of regulatory process, no matter how long they are, cannot be counted as proteins. So, how many of the 3.15×10^{57} amino acid chains generated by our calculations actually have those characteristics?

In a study performed at "The Research Center for Probabilistic Calculations in Biology," in the US, words consisting of an increasing number of letters—2, 3, 4, 5, ..., one after another— were written by randomly choosing letters from the alphabet. Then, the meaningful words which resulted were counted, one by one, and their sum was compared with the total number of outcomes—both meaningful and nonsense words—to reach a statistical conclusion: the probability of a meaningful word occurring as the result of random draws from a 20-letter "amino acid alphabet"—that is, the odds of obtaining a protein that could actually take part in a structure or function—was $P = (1/4)^n$, where P is the probability of the chance occurrence of the protein, and n is the number of amino acids in a given protein.

Thus, the probability of a protein chain containing a small number of amino acids—for example, 100—forming by chance would be defined by the equation, $P = (1/4)^{100}$, yielding one in 6.22×10^{61}. Then, to calculate how many such protein molecules would have formed over 5 billion years, the number 3.2×10^{57} would have to be divided by 6.22×10^{61}. The result is approximately 0.00005—meaning that the chance formation of even only one protein molecule which would be morphologically or functionally useful is impossible (it has a zero probability), and this can be stated with mathematical certainty.

Alexander G. Cairns-Smith, of Glasgow University, describes this zero probability in the following comment: "If the entire world was full of amino acids 5 billion years back, and there was nothing else present, and even if those amino acids made 10 bonds every second, the probability of occurrence of only one protein molecule, for instance, the probability of only one insulin molecule, forming by chance would be as small as zero."[170] The following example, by George Gamow, clarifies the subject even more: Get a glass of water and put it on your table. Have you ever thought about how this refreshing water could be a source of danger at all? The H_2O molecules composing water are always in motion, just like all other fluid molecules. Each molecule may tend to move in any direction in a disorderly manner (indeed this disorder is such an order that we have not been able successful in measuring it). For those molecules (x), each moving in various directions, it is probable as $P = 1/10x$ for all of them to start going in the same direction. For instance, if all of the water molecules in this glass move upward by chance, the water will become faster than a missile, while it's standing still on our table, and it will jump up toward the ceiling like a bullet. Mathematically, even the occurrence of this is more probable than the formation of only one protein molecule by chance; so far, no one has ever happened to observe such a case that our reasoning denies, and as long as the world exists, no one will ever see it.[171]

A similar calculation could be done for a small protein consisting of 100 amino acids. Those 100 amino acids could be sequenced by chance in 10^{158} different ways, with only one of those ways yielding the required protein molecule. If all 10^{80} atoms in the universe could be used to form a protein molecule having 100 amino acids, then 10^{78} groups of 100 units could emerge at any time. Each time, if the combination we obtained were not the desired one, then we would simply put all 100 amino acids we drew "back into the bag," as it were, and then draw 100 successive amino acids once again. Supposing that we made one billion (10^9) draws in a single second, and taking the age of the universe to be 30 billion years (10^{18} seconds), the number of these types of combinations would reach 10^{105} (10^{78} x 10^9 x 10^{78}). This means that the chance of one of those proteins being the desired one is a minuscule number—1 in 10^{53} ($10^{158}/10^{105}$). However, most proteins present in living organisms actually consist of more than 400 amino acids, reducing the odds even further.[172]

James F. Coppedge, in his book, *Evolution: Possible or Impossible?*, gives the broadest information about probabilistic calculations.[173] Critical information, given as quotations by researchers, such as Harold J. Morowitz, in Chapters 1, 4 and 6 of that book, under the subtitle, *"Molecular Biology and the Laws of Chance in Nontechnical Language,"* attributes the notions of chance, coincidence, and "accidents" to a completely invalid historical argument. Coppedge also did several probabilistic calculations, all showing the extreme improbability of life occurring by chance. According to him, in order for protein formation to occur in primordial Earth's conditions, where such formation was extremely unlikely to happen in the first place— even if we suppose that all the conditions were suitable, such as the rate of reactions forming amino acid chains being one-third of a ten-million-billionth of a second (note that this concession means that 150 thousand trillion amino acids could actually be made in a single second at a normal speed), we arrive at a probability value of 1 in 10^{287} for one protein forming from a chance sequence of amino acids. For the minimum set of 239 protein molecules to have even the smallest theoretical life, the probability of chance formation is 1 in $10^{119.879}$. Surely, this defines the impossible.

In turn, according to Frank B. Salisbury's calculation, the probability of the chance formation of a protein composed of 1,500 amino acids is 1 in 10^{450}. If a trial were performed in a billionth of a second, and if 10^{80} (the number of atoms in the universe) amino acids were entered into this trial; and third, if we assumed this process to be going on for 30 billion years (10^{18} seconds)—then the total number of successful trials over time would be 10^{107}. That is clearly a much smaller result than 10^{450}.[174]

Salisbury also clearly points out that genes appear to be too unique to have occurred by chance. In his opinion, even if genes had managed to come into existence by chance, a certain enzyme would have been needed at some point. The evolutionists claim that this early enzyme appeared as a result of chance mutations in existing genes. Yet performing a calculation for genes to form by chance, Salisbury arrives at a conclusion that will make the reader smile: if we generously assume the number of planets to be 10^{20}, and we further assume that each is replete with oceans consisting of small DNA genes of 1,000 nucleotides in length, which replicate a million times a second, with a mutation occurring each time, the odds of getting the desired result is 1 in 10^{415}. According to him, then, it is simply too improbable that natural selection and chance could have formed life if the Earth is only 4 billion (4×10^6) years old, and he deems that this poses a real dilemma since natural selection and chance need something to operate on.

On the probabilistic calculation of the occurrence of life by chance, Yockey states that a small polypeptide molecule containing 49 amino acids could emerge from among the amino acids that have biological activity in pure water in 10^9 years.[175] However, even the single cell, which could be a model for the simplest hypothetical living being has 256 proteins.

Insulin protein, one of the smallest protein molecules, is composed of 51 amino acids and has a molecular weight of about 6,000 Da. It is a very important hormone for the regulation of glucose utilization in the body. To calculate the chance occurrence of such an orderly chain, arranged by links between amino acids at particular points, arising to form the insulin protein, we need to calculate the huge number of 20^{51}. The number obtained would be so big that it could not fit into billions of multiples of the lifetime of the universe. In turn, the proinsulin molecule, which actually forms insulin and is more complex than insulin, is composed of varying numbers of amino acids, from 81 to 86, in various groups of animals. Supposing that a particular proinsulin molecule had 84 amino acids on average, the probability of the chance occurrence of one proinsulin molecule being made from 20 types of amino acids would be 1 in 20^{84} or 10^{109}. Thus, though reading this number, with 109 zeroes, is not easy, it is even harder to claim that proinsulin can be formed by chance.

As innumerable experiments performed so far indicate, life does not occur spontaneously anywhere. The mathematical approach, which uses probabilistic calculations, also eliminates the possibility of life having arisen by chance, either terrestrially or extraterrestrially. This leaves only the one option: the Creator. In other words, life had to be created by a One Who has infinite knowledge and power to design and organize everything for every single creature, from atoms and cells to galaxies. Unfortunately, even with the slightest probabilities of chance occurrences, there will always be those who say that such improbable events still have a chance of occurring, no matter how slim the odds.

Darwin believed that given enough time, small changes accumulating over time could account for the transformation of one species into another. However, since all of those changes in living mechanisms would be chance occurrences, without a purpose or goal, could one reasonably expect that they could be charged with the formation of all of the highly complex, well-ordered, precisely functioning organisms that make up the plant and animal kingdoms? Darwin staked his professional reputation on this very expectation. He proclaimed that it was all a matter of probability. According to evolutionary thinking, the principles of probabilistic calculations cannot preclude a possibility from occurring. Even for the most reasonable evolutionists, then, the chance occurrence of a thing always exists, despite the fact that it has never happened in the past, and that the probability of it occurring in the future is statistically improbable. Yet, according to probability theory, although the chance of getting "heads" every time in one million times a coin is tossed up is extremely tiny, it is deemed possible, statistically speaking.

Furthermore, Darwinists continue to argue that time is on their side. They point to the age of the Earth, five billion years, and claim that it is surely a sufficient length of time for chance mutations to have added up to significant changes. No one would deny that 5 billion years is a long stretch of time, but is it long enough to account for the chance evolution of the whole complex of life, in all its myriad forms? Mathematicians would answer this question with an unequivocal, "No!" Some of the world's greatest mathematicians, in fact, have deliberated on and played with evolutionary claims, attempting to match time spans to mutation frequencies and the formation of organized living systems; but in the end, they always end such endeavors by throwing up their hands in complete disbelief of evolution. According to all their calculations, the statistical probability that ordered life emerged from chance occurrence and accidental arrangements of mutations is virtually zero. In the world of statistics, events whose probability lies within the range of $1/10^{30}$ to $1/10^{50}$ are deemed impossible.

Let us examine a simple single-celled organism and take it as a gauge. A living cell is an astonishingly complex mechanism consisting of thousands of organelles and myriads of diverse chemicals, all finely organized and functioning in a mutually beneficial and orderly fashion.

Even the staunch advocate of evolution, Carl Sagan, points out that in terms of information alone, it is estimated that a one-cell bacterium of *Escherichia coli* contains one trillion bytes of information. About 100 molecules are synthesized by enzymes every second, and they become divisible in 10 minutes. It has been estimated that this amount could be compared to 100 million pages of Encyclopedia Britannica.[176]

Jeremy Rifkin mentions that even a tiny, one-cell organism is certainly something to contend with, and after explaining Simpson's opinion in a way which makes the portion of the evolutionary trip leading up to the simplest one-cell living mechanism seem as impressive as the rest of the evolutionary journey combined[1], he reports that above the level of the virus, the simplest fully living unit is almost incredibly complex. It has become commonplace to speak of evolution from amoeba to man, as if the amoeba were the simple beginning of the process. On the contrary, if, as must almost necessarily be true, life arose as a simple molecular system, the progression from this state to that of the amoeba is at least as great as from amoeba to man.[177]

Let's follow Rifkin's observations a little further:

> Apparently, the mathematical odds more than agree with Simpson's analyses. In fact, according to the odds, the one-cell organism is so complex that the likelihood of its coming together by sheer accident and chance is computed to be around $1/10^{78436}$. Remember, nonpossibility, according to the statisticians, is found in the range of $1/10^{30}$ to $1/10^{50}$. Needless to say, the odds of a single-cell organism ever occurring by chance mutations are so far out of the ball park as to be unworthy of even being considered on a statistical basis. When one moves from the single-cell organism to higher, even more complex forms of life, the statistical probability shifts from ridiculous to preposterous. Huxley, for example, computed the probability of the emergence of the horse as one in $10^{3,000,000}$.
>
> Albert Szent-Gyorgyi, a Nobel prizewinning biochemist, says he can no longer buy the Darwinian interpretation of evolution. Regarding the supposition that random mutations over time do indeed account for the accidental formation of all living things, Szent-Gyorgyi says that he simply cannot accept, "the usual answer . . . that there was plenty of time to try everything." This eminent scientist admits: "I could never accept this answer. Random shuttling of bricks will never build a castle or Greek temple, however long the available time."
>
> A conference was convened at the Wister Institute of Anatomy and Biology in Philadelphia to address the question of the mathematical probability of evolutionary theory. In atten-

dance were some of the world's prominent mathematicians and biologists. The latter group was not pleased with what the former group had to say. After making all their computations, the mathematicians concluded that there was not enough time in the entire universe to account for the statistical probability of life forming spontaneously by chance mutation.

As to whether chance mutations, working through natural selection, can, over a sufficient period of time, produce complex living systems, computer scientist Dr. Marcel Schutzenberger, of the University of Paris, concludes: "We believe that it is not conceivable. In fact if we try to simulate such a situation by making changes randomly at the typographic level . . . on computer programs we find that we have no chance (i.e. less than $1/10^{1,000}$) even to see what the modified program would compute; it just jams. It is our contention that if "random" is given a serious and crucial interpretation from a probabilistic point of view, the randomness postulate is highly implausible and that an adequate scientific theory of evolution must await the discovery and elucidation of new natural laws."

The findings of the mathematicians were upsetting. After all, evolutionary doctrine owes its very existence to probability theory. For nearly a century, biologists have been preaching that random mutations can account for meaningful structural organization and reorganization over a long period of time; and they have used the notion of statistical probability to make their case. Now some of the world's leading mathematicians say there just isn't enough time, statistically speaking, to account for complex sophisticated living systems by the accidental shifting and rearrangement of genetic mutations. Their conclusion serves well as both a summation of, and a final epitaph to, the neo-Darwinian synthesis: "Thus, to conclude, we believe that there is a considerable gap in the neo-Darwinian theory of evolution, and we believe this gap to be of such a nature that it cannot be bridged within the current conception of biology."[178]

9

Toward a Model of Creation

TOWARD A MODEL OF CREATION

W hether conditional or open to different alternatives, our opinion—arrived at by taking all of the ideas in the field of study into consideration, and depending on the present information or data—does not need to harm our connection with our Creator. The points on which the present positive sciences shed light might actually describe the creation process of the universe, the Milky Way and the Earth very closely—or they might not exactly. The important thing is to comprehend that those processes are indeed a "cloak" for our Creator's knowledge and power, in that refinement, perfection and greatness are all abundantly evident in the creation mechanism.

According to modern knowledge and scientific advances, if the phenomenon we call "life" is indeed only present on Earth, then we could say that the last to be created of all the dynamic pieces (subsystems) of our planet is the biosphere. The preparation of Earth in this way, to make it suitable for life, moving step by step through all the stages mentioned above, starting from the Big Bang—just like completing an elaborate piece of art very slowly from hundreds of building blocks—is judged possible only as the manifestation of infinite knowledge and power. Our estimations—which use some information, to obtain some clues, for some of the processes—are merely our attempts to shed light on the chain of causes which veil the divine creation process. In this way, thinking about the various possible ways by which creation might have happened, using some of the presently available evidence—without exceeding our limits of understanding before God's inestimable act of creation—should only cause a believer to strengthen his or her faith in God. However, saying that the creation certainly happened in this way would definitely be both incorrect and presumptuous. It is not difficult for our Lord, in His infinite knowledge and power, to exhibit different ways of creation that no one could ever have thought of. A human being, with only limited knowledge and curiosity for research, is able to discover but some clues in the reflections of reality hidden behind hundreds of veils. Those discoveries lead humanity not to a life phenomenon that is a mere game of chance—but to the Creator, Who has Infinite Mercy and Compassion.

If God wills, He can create or destroy all of creation in an instant. Both creation and eradication are just as easy (to the same degree) for Him; nothing is beyond His knowledge and power. No believer has a right to object, in any event, as He makes use of His property according to His will. However, since this world is a place to test human beings, God uses causes in both processes—creation and destruction—as a veil to His greatness and magnificence. He has provided some principles and laws for us to use in searching to uncover the mysteries of creation, thereby allowing us to make connections between the cause and effect relations of certain events. In addition, God allows us and wills for us to look into the universe, so that we may think about creation and find Him through the benefit of our talents, which He has given us, such as intelligence and curiosity.

If He had willed, He could have destroyed everything in a period of time as short as the blink of an eye—and He could have recreated everything in the exactly same way again. Also, He could have written His name in the stars, and He could have clearly stamped His name on the faces of everyone. However, in that case, everyone would have to believe in God—as the truth of the test of this life reveals—and such faith would not be as worthy, since the human being's limited willpower would have no bearing, and we would simply be forced to believe.

However, what is most worthy before God is for human beings to perceive the stamp of creation which is hidden behind material causes by observing the perfection, harmony and beauty of creatures—ornamented with such fine peculiarities and talents—using the intelligence and the limited willpower given to them.

In other words, the cause and effect chain which connects the creation process, as we try to discover its mysteries through various fields of study, was only placed there for the benefit of our will-power and choice—not to deny it.

Therefore, when we consider the aspects of the subject matter which are commonly accepted today—starting from the Big Bang and moving through the phases which were explained briefly above—we should keep in mind that each of the astrophysical and physico-chemical processes which we have been able to determine relating to the creation processes of atoms, molecules, galaxies, supernovas, suns, stars, the Milky Way galaxy, the solar system, and the Earth all veil the divine act of creation. Attributing the creation process completely to cause and effect relationships (i.e., absolute determinism) is totally different from seeing the Ever-Able Artist Who applies His unlimited willpower to generate causes as veils over His splendor and magnificence. Instead of accepting the laws which are present in the universe from a strictly deterministic point of view, one should keep in mind that opening a door to the human intellect—and so not rejecting completely the phenomena which are linked to causal rela-tionships—is a necessity of our being tested in this life. In other words, it is sometimes possible for humans slightly and partially to lift this veil of causes, with our limited mind and curiosity, to see a determinism which is dependent on conditions. We may even make circumscribed interventions in some key processes from time to time, as a necessary concomitant of the vicegerency position given to humankind (sometimes having to endure the results of our own interventions, as in cloning and playing with the genes of living organisms without having reflected sufficiently deeply first).

Certain phases in the earth's being made a suitable place for life may have had similarities with evolutionist arguments; for "great minds think alike." Nevertheless, a process that has been set forth with a detailed plan and delicate calculations in accordance with the preferences of the Creator's infinite knowledge and willpower rejects chance completely. The creation of human beings and animals had to be preceded by the creation of the atmosphere and water. The creation of such a magnificent and marvelous molecule as chlorophyll can be estimated, as can the presence of free oxygen in the atmosphere. While the creation of the chlorophyll molecule itself required a very significant source of energy, the Sun, to be present in order to serve life, there was no opportunity for the Sun's radiation to be used in any synthesis reaction before the creation of chlorophyll. Nonetheless, infinite knowledge and power is essential in order for the chlorophyll molecule to be given in the service of life as an amazing energy transformer. This is because no other kind of power, possibility, chance, or nature could have formed chlorophyll as such a perfect and unique structure.

Metabolic processes might have been changed by the creation of aerobic respiration which released sixteen times more energy than fermentation (for example, the "Pasteur effect" might have started with 1% oxygen compared to the present ratio). Two different directions for creation could have been anticipated as a result of the initiation of respiration as either a heterotroph (a "consumer" of carbonaceous organic compounds) from the animal kingdom, or as an autotroph (a "producer" of carbonaceous organic compounds which uses sunlight and consumes minerals) from the plant kingdom. On the other hand, when it is logically considered, the creation of plants, which are given the ability to synthesize their own food in advance (by means of the presence of chlorophyll) should come first, and then the creation of animals that are in need of plants because they cannot synthesize their own food should follow.

From this point of view alone, molecular oxygen could be considered to be the basis of life. And yet, it is not. Molecular oxygen is only useful to a metabolism that uses a great amount of oxygen (like the oxidation of pyruvate, which is a product of the breakdown of glucose). Conversely, however, molecular oxygen is also a poison for all organisms, which do not have the protective enzymes required to reduce the effects of the damaging waste products. This means that organisms which are described as "primitive" by some scientists actually represent amazing and very complex biochemical laboratories. For this reason, we can conclude that the essential considerations of many stages (including the random synthesis of the first molecules, the formation of coacervates, and then of the first molecules), which are supposed to come one after the other according to evolutionary theory, remain clouded and hypothetical.

IF THERE IS A WORK OF ART, THEN THERE IS AN ARTIST

Some scientists believe that the explanation of the universe and life should be based solely on natural factors. However, the foundation of this belief is a preconception of the universe and life as being the production of merely physical powers. But what if this was not actually the case? Even when we see a pair of eyeglasses, we can make a judgment that they are not the product of physical powers only; rather, they are made by an intelligent and skillful optician. Nonetheless, life is thousands of times more complex than a pair of eyeglasses. Thus, we come to the conclusion that life has to be created by an intelligent and talented Power. Here, the crucial requirement is to succeed in evaluating the scientific evidence without prejudgments, as far as possible. However, Darwinists argue that science cannot acknowledge a supernatural power—though the majority of scientists, in fact, accepted a creating power, God, until the middle of the nineteenth century. It seems that the claim that science has to be materialistic arose after Darwin, but this claim is increasingly contradicted by scientific evidence. Undoubtedly, the reason why a field of science like biology has been distorted and made an instrument of materialism is that it arrives at a common point with Marxist and atheistic views, since evolutionists, Marxists and atheists all look at the subject matter (biology) through ideological eyeglasses, and that ideology—which reflects both the fundamental thinking of certain interest groups and their own worldview—is made to appear strong through intense propaganda in the mass media which supports them.

One of the most important reasons why evolution spread so rapidly as an idea for 150 years is that evolutionists were able to say whatever they wanted, in the absence of rivals, until about fifty years ago; strong voices opposed to evolutionary scenarios did not emerge for almost a hundred years. In particular, ideas like "social Darwinism" provided opportunities for applying evolutionary notions to society, and scientists who believed in the Creator were psychologically oppressed or effectively silenced so that they would not, and could not, speak out against findings reported in science magazines. In some countries, like Turkey, they were directly suppressed through harsh policies—all of which prepared an ideal environment for evolutionary theory to spread easily. Another important factor which made the evolutionists' job easier, and allowed evolutionary theory to gain wide acceptance, was that in the long-standing struggle of contradictions between science and religion in the West, Christianity had not been sufficiently able to withstand discoveries and debate, and thus scientists had been forced to keep their distance from the Church, essentially since the Middle Ages.

On the other hand, the very firm tenets of Islam concerning the sciences do not give the opportunity for controversy, or conflict, between science and religion. However, as a result of those studying religion simply abandoning the sciences, and those studying the sciences being deprived of religious education, there did occur an artificial separation between science and religion, even in Islam, and enmity arose as a result. This situation was utilized expertly by atheist and materialist special interest groups who sought to dominate the education system. There was even intense propaganda sponsored by individuals of a certain mentality which aimed to get people to associate completely negative and inverted images—such as considerations of superstition, bigotry, fanaticism, and reactionism—with religion. In the absence of scientists who were knowledgeable both in the sciences and religion—and given that those who were knowledgeable in both fields were often overwhelmed by the mass media's promotion of evolutionist propaganda—the stage was set for advocates to present evolution as if it were a proven science. Ultimately, science should not have been used to render this materialistic interpretation of life, but rather, to provide a true explanation of it. Some people's philosophical persuasions were disturbed, but only absolutely true evidence should be followed, and the information coming from religious sources should never be approached in such a way as to reject it with prejudice.

Today, if you meet with ordinary people and discuss their ideas about evolution, you will see that the majority of them do not believe in it. However, many of them do not have real knowledge about science, and they rely on traditional cultural and religious teachings for their worldview. Conversely, the majority of evolutionists have acquired their worldview after a certain level of education caused a big rupture in their belief systems. However, it should actually be totally the opposite, for science education should bring people to faith—not away from it—and it should teach us to read the book of the universe correctly. Yet, reversing this picture will only be possible through the efforts of a new generation of young people, who will give their utmost efforts to making science and religion embrace each other, and who will, with the best of intentions, succeed in uniting their minds and souls.

In many ways, this emerging point of view, regarding the need to unite science and faith, which has been brought about by both the subject of evolution and general conditions around the world—and which is parallel to a similar revival in the west—is permeating Muslim countries like Turkey at this time in history. Those who unfairly blame Muslims for being unprogressive and "enemies of science" are now themselves beginning to seem "unprogressive and bigoted." In fact, except for some strict atheists, many scientists in the west are now actively questioning Darwinism and the general foundations of evolutionary theory. They might not be able to bring out supportive scientific evidence from the Bible, but they shake evolutionary theory at its roots by means of very strong scientific and mathematical evidence, and we can say that they have at least partially defeated evolutionary dogma. The biggest advantage which Muslims have in this regard is that the divine revelation which Muslims follow, the Holy Qur'an, is uncorrupted. Should scientists who currently keep themselves away from this amazing resource, the Qur'an—the interpreter of the book of universe—actually approach its verses concerning creation with reason, and without prejudice, they would indeed arrive at ever more critical conclusions, and even wider-reaching opinions.

As a matter of fact, the struggle between belief and disbelief has been going on since the first human being emerged, and it will continue until doomsday. Therefore, no matter what one proposes as evidence, or what type of logical explanations one offers, or how many exemplary phenomena one shows to those who take to the road in the name of denying God— the choice of belief versus unbelief being the essence of our test here in this world—some will always find a way to embrace disbelief. We cannot avoid this, and we also cannot ignore the reality that the subject matter of evolution has a dimension which connects with predestiny, so that our desire to search for Truth in this field of discovery is itself only inspired by God. Thus, even though we can clearly prove a divine origin for life, with countless forms of evidence, God's letting people perceive Him in their hearts only occurs as He ordains. Our duty, then, is only to put out for everyone to see clearly the distortions of science introduced to the public in the name of unbelief in God. In democratic systems, everyone has the freedom to stand up for all kinds of thoughts, to take these up seriously, and to explain them to others. So we also have a right, the most natural right, to mention our belief in God upon finding an opportunity to do so. We showed in various ways above how impositions in the name of science became instruments of distortion, forgery, and misinterpretation. Today, evolutionists have reached a point where they have started hardening the debate, as they see that their place on center-stage is disappearing, and the numbers of adherents to their ideology is gradually decreasing. In some countries, like Turkey, to extend the example of my own background experience, evolutionists have adopted a particular stance, in virtually every scientific subject, which effectively obstructs the rights of their opponents. However, in the wake of recent technological advances in many countries, particularly in the US, and the effect of the Internet, which distributes without boundaries all kinds of information to anyone who cares to find it, all indicators are that the evolutionist position will slowly weaken.

This does not mean, however, that evolution as a concept will entirely vanish or that it will become completely irrelevant. There will always be adherents and believers in evolution as dogma, as a belief system, because if they are as atheists or materialists, all humans need to respond to the search inside themselves, and to connect to some kind of faith in something. Thus, even if it is not fully satisfying, many people will choose to believe in evolution, and so to experience the freedom of deception in lieu of the duty to worship God.

Of course, the notables and followers of evolution will continue to evaluate all types of discoveries and new findings in biology from their own perspective. They will feel obliged to find some rationalization in each new discovery, such as the Human Genome Project, stem cell treatment methods, and genetic improvement and treatment techniques. Actually, they should not be blamed for upholding a worldview in which every event is witnessed as a basic reflection of their beliefs. For just as those who believe in God see the manifestations of the names of God on the wing of an insect and in the eye of a butterfly, evolutionists look for evolutionary mechanisms in the very same entities, and make their interpretations accordingly.

The important thing is not to distort science and not to lie. The right of interpretation is surely a necessity, and privilege, of democracy and independence. Up to now, the advocates of evolutionary theory have used this freedom in all ways, while they have accused those who believe in creation of being unscientific and reactionary. They are not even able to tolerate the teaching of both systems of thought together in schools, for they insist that evolution is "scientific," and they demand that the teaching of creation be cancelled entirely so that evolution can be taught exclusively.

Of course, in order for their demands to be satisfied, they would obviously first have to clarify the definition of "scientific;" then, they would have to answer the questions raised above, one by one.

In fact, the biggest problem with Darwinism is that it sees a perfect universe and wonderful ecosystems, and the entire world of living beings, as the work of blind chance. Yet a belief system conditioned by a lack of supervision, purpose, and use, which is essentially a wild, brutal struggle, as opposed to a place of wisdom, meaning, planning, and beauty upon all creatures, should be ready to state outright what, exactly, it offers humanity.

Proponents should also explain in terms of biology how an organ (like a fin, wing, heart, kidney, and so on), never seen before and without a prototype, somehow emerged in a group of animals at the exact right place and in the most ideal way; where were their plans and projects drawn; who wished it to be formed in such a fashion. They would also have to answer the question about which biochemist's instructions these perfect cells, each operating like a factory, would be following in their functioning.

It is important to note that Darwinism may address the issue of how biological structures already existing might have undergone some small changes; for example, it can propose an explanation for how the small differences in the beaks of finches on the Galapagos Islands first appeared. But questions about how those birds came into existence in the first place; or about how sophisticated features, such as the feathers and the wings of these birds, have assumed their present forms; or about how really intricate, delicate organs and systems, where countless components work harmoniously, like in the operation of the brain and eye, and in blood clotting, came to existence—none of these can be answered by Darwinism, as each of these entails such complexity that the organ, feature, or system as a whole can only function when every component is fully operational and free of defects. The most logical way of explaining the origin of those organs and functions is to acknowledge the intervention of a supernatural Creator Who is conscious and Who has infinite knowledge and power, and evolutionists will never be able to "get rid of this trouble."

In the past, prior to many scientific developments and revolutions, certain fanatics insistently supported outdated theories in a similar fashion. But, after some time, their mistaken notions collapsed in the face of the increasingly undeniable evidence which was proposed by more objective scientists. In a similar manner, the evolutionary idea is bound to capitulate before the overwhelming and convincing discoveries of scientists whose objectivity is rooted in the fact that their hearts and minds are united, individuals who can read the book of universe externally, through meticulous observation, and internally, through sincere contemplation and whose intentions and actions are therefore clear and unobstructed.

On the other hand, the growing number of individuals who successfully combine science and faith in God will not mean that we will see an end to the conflict between belief and disbelief. Having started with the first human being, this struggle will go on until doomsday. Even if Darwinism were fully abandoned today, we should expect another ideology, philosophical school or worldview—wrapped up as a "scientific taboo"—to be introduced to the public in the name of denial and unbelief.

Our efforts to disprove evolution are not rooted in a rejection of the materialistic and atheistic worldview that the evolutionary idea aims to bring about; rather, it is because of the fact that evolution has been argued to be "a proven law," and "a truth that has to be believed." Yet in highlighting their own beliefs and values, those who believe in God have been labeled "outdated" and "reactionary." Furthermore, we should point out that there is actually no obligation or necessity to introduce a "creation model," for creation is a miracle hidden behind veils of causality, and explaining miracles within the limits of the normal laws of nature is not something which is possible. In fact, when we look at things from this perspective, many of us make the mistake of expecting miracles to happen in an obvious way. We expect only events of a certain magnitude, such as a child surviving a fall from a 100-storey skyscraper, or a tree stepping out of the Earth and moving on its own. Yet these are such transparent and obvious events that the mind would simply be dazzled, and the intellect rendered helpless.

However, innumerable, astonishing, perfect processes occur constantly in our bodies and in other living beings—the formation of an image on the retina of the eye; the perception of sensation in our brain; blood being filtered in our kidneys; transmissions along our neural pathways; the contractions of our muscles and the movement of our intricate joints—each is created and executed with wisdom, each is an artistic structure, and each is a miracle. However, if a particular event occurs frequently, after some time the human mind starts seeing it as being common and normal. Thus, even the most amazing phenomena come to be taken for granted. There are millions of births happening routinely, for example, and we consider these to be very simple events. However, when we carefully and objectively examine the 280-day process which passes, on average, from the meeting of the sperm and egg to the birth of a human baby, and we further analyze the development of fetal tissues and organs, day by day, then we will be compelled to assert the miracle of every single birth. If we could imagine this nine-month process being fast-forwarded to a single half-hour period (so that a baby would be born half an hour after fertilization), then perhaps we would better comprehend this miracle. But we are not able to see the miraculous aspects of phenomena hidden behind the veils of apparent causes (such as DNA, genes, and molecular, biochemical, physical, and metabolic events) that are put before us as part of our test of belief, and which occur in subtle or repetitive ways over a wide span of time.

On the other hand, in order for us to propose, in detail, any kind of system or mechanism as a possible "creation model," we would require as much knowledge and power as our Creator has. This is because succeeding in performing an incomparable action, such as giving or creating life, necessitates being matchless and unique—but the attributes of infinite knowledge and power are only associated with God. As human beings, we have neither witnessed creation, nor do we have the ability to apprehend such a miracle. Our created brains and hearts are not supposed to see or perceive the Creator in person, using the senses given to us. Rather, we believe in God solely after accepting the reality that "creation must have an Originator," as the mind, heart, and conscience work in harmony with the senses. Those who are created can neither interfere with the work of the Creator, nor ever understand how such artwork is actually executed. We can only try to comprehend some aspects, to a certain extent, using as much evidence as our mind can grasp—and try to strengthen our faith.

We may ascertain this point better with the following example. Let us assume, for the sake of argument that hundreds of complex computers in a huge computer laboratory are talking to each other within the limits of the software and hardware installed and searching for answers to questions about how they came to this facility in the first place, and how they were built. What those computers "say" to one another, what they claim or discover, and all their brilliant ideas—none of these can ever go beyond what their programs allow them to achieve. They may discuss their hard disks, RAM, processing systems, keyboards, drives, and video cards. However, they will never be able to know the kind of person who made them—the computer engineer—as to which attributes this person possessed, or this person's real character.

Just like this example, as computers cannot get to know the engineer who designed them, we cannot apprehend the Essence of our Creator nor can we totally understand how He created us, nor can we ever propose a comprehensive model which displays one-to-one correspondence with reality. Simply put, we can neither conceive nor say more than what God has taught us and allows us to say.

IN BETWEEN RELIGION AND SCIENCE

In the recent past, many people used to propose objections against Darwinism based solely on religious grounds. For their part, advocates of evolutionary theory used to claim that science was only on their side. However, scientific findings which have been obtained since the last quarter of the twentieth century have reversed the picture. Today, our objections are not because of the things we do not know, but rather because of the things we do know. Now, those who seem dogmatic are the Darwinists, for the world of science provides them with ample evidence that life has been created with a plan and program, but they deny this evidence out of hand because of their philosophical and ideological worldviews.

In any event, why would it hurt if an idea, doctrine or thought system were inspired by religious sensitivities? The important point is whether or not the things one says contradict the intellect, reason and real scientific findings. Religion is vital for human beings, and humans cannot live comfortably within a duality: we cannot be content in a world where the natural need for faith which is in our hearts and souls is split from the efforts and determination of our minds and our science. Believers cannot, and should not, compromise their faith in God, His names, and His attributes, nor associate these names and attributes with mere causes, chance and rambling atoms. Those who believe in God cannot accept the notion of a deity which does not control everything, or which only exerts partial command over the created world, from atoms to galaxies, or which would be uninformed in regard to the finest details of the wing of insect, or which would be unaware of what has already come to pass. Evolutionary theory, on the other hand, tries to embody all the characteristics of God, essentially while failing to fulfill any of the deeper spiritual needs of believers. This is a complete contradiction which must be seen for what it is—our aim here is not, as mentioned earlier many times, to oppose science.

Yet the fact that we are not able to say anything about the first creation should not cause us to abandon the causal aspects of creation. On the contrary, each new piece of information that scientists uncover, each new beauty which is exposed, should increase the believer's astonishment and admiration. Even though it is not possible for us to display the initial creation with all of its details, the perfectly functioning processes that we witness by the millions every day in the births of plants, animals and humans, and in the organs and physiological processes of living things, are all waiting to be discovered as evidence in favor of belief in God.

Too many scientists have spent too much time, energy, and effort in vain, for one and a half centuries, denying God on the basis of Darwin's evolution hypothesis. However, if the efforts of scientists had been directed instead to the countless genetic diseases, or to cancer research, or to the environmental problems which are now in humanity's hands, remedies for most of those problems would have already been found, and countless improvements to the human condition would have been achieved by now. What kind of benefit does the scientific community obtain by talking out of place concerning the first creation, and continuously interpreting it with the aim of denying divinity? Furthermore, since the negativity of chance, meaninglessness, deficiency and failure will be seen when looking at nature from the evolutionists' point of view, the resulting perspective will have an obstructive influence on scientific improvements. In contrast, objective scientists, who uphold a worldview wherein science and faith are compatible, would never see deficiencies, defects or ugliness in creation—rather, they would simply search for the wisdom behind every event, and all scientific studies would only increase their faith.

In the Holy Qur'an, after pointing out evidence from the book of nature and mentioning various events, many verses encourage people to think and search with questions like, *"Don't they think?"*; *"Don't they contemplate?"*; *"Don't they reflect?"*; or *"How can you deny?"* Thus, as seen above, faith in One God calls us to search, to work, and to benefit humanity. However, a great many efforts have been directed to "uncovering" the essentials of the first creation—engendering ventures which have not profited anyone—as if humanity did not have any other problems to attend to. What would happen if, without any veils whatsoever and without connecting them with causes, God had actually shown us how He had created the first living beings, the first ancestors of each species, and the first humans? Those who believe in God would already believe in God, even with the veils of causality—and when there were no longer any veils, the value of believing in the unseen, the value of the test of this life, would simply diminish. Furthermore, while more people would believe, there would, no doubt, still be unbelievers. However, we were created, we are being tested, and we have not determined any of the conditions of this test ourselves. God does everything in the way He wishes to do it; He creates everything whenever He wishes, and He destroys things whenever He wills to do so. Rather than preventing anyone from researching or studying, the refinement and beauty in the artwork of God's creation directs us to see beyond the artifice on the horizon of bewilderment and thus, to increase our faith.

Regarding how religious beliefs deal with the debate on evolution, we should first point out some of the differences in the perspectives of Christianity and Islam. In the holy books, revealed through various prophets at different periods in human history, God informed people about Himself according to their level of comprehension, their knowledge, and their cultural accumulations, and according to their needs at the time they were living; in short, God guided people with examples that minds could grasp at that point in history. Some of the information given was rather obvious, some was made easier to understand with symbols and analogies, and some things could only be understood based on the explanations and clarifications provided by the prophets. For this reason, a special way of commenting or interpreting the Qur'an, called *tafsir*—exegesis—was developed to relate the Divine will in the best possible way so that the information given would fit the understanding of the time.

The failure of the church in interpreting the Bible as it was supposed to be played an important role in the discord between the church and science during the Middle Ages. For example, debates around the revolution of the Earth, the creation of the universe in six days, and the idea that one of Eve's ribs was missing, originated in misinterpretations of the relevant verses in the Bible.

Along with the weakening of the Roman Catholic Church's authority, the scientists' readings of the book of the universe were increasingly deemed to contradict the deductions made out of the word of the Bible. In fact, if we look at even one example through the evolutionist lens, we can easily grasp how conflicts arose. For instance, with regard to the belief that the universe was created in six days, Bible literalists insisted that "six days" referred to six 24-hour, worldly days. Yet advances made in geological and paleontological research indicated that the Earth had been formed over very long periods of time, measured in thousands or billions of years, which could not be equated with the worldly sense of a "day." As a result, scientists found themselves forced to choose between believing in field observations or Biblical interpreters. Thus, one discovery at a time, the fight between science and religion took shape.

The subject of creation taking place in "six days" is found in the Qur'an. However, the six days in the Qur'an are not defined as 24-hour days, such as those we have on Earth. References in other verses of different chapters, regarding the possible length of those "days," point to a "day" being perhaps as long as 1,000, or even 50,000 years. Of course, the time span which we call "one day" signals one full revolution of our Earth upon its axis. Yet when we define time from another reference point, the length of one day, which depends on the rotational movement of a given astral body— for example, Jupiter, or a meteor, or a planet in very distant galaxy— will be very different. Further, if we take the motions of meteors as a case in point, we could also think about the various lengths of time required in terms of the velocities of angels or other spiritual creatures. The fact that such matters remain undefined certainly makes it easier to interpret the Holy Qur'an, since those six days may not even necessarily be equal to each other. Most importantly, we may think of those six days being six different "phases" of creation—for instance, the creation of atoms, molecules, galaxies, the solar system, the Earth, and the biosphere. On the other hand, from the geological point of view, we may consider them as six geological periods like the Precambrian, Cambrian, Paleozoic, Mesozoic and Cenozoic. In terms of biology, we may imagine yet another scheme of "six days" in the creation of the Earth, oceans, atmosphere, green plants, animals, and human beings, in that order. In effect, such Qur'anic verses with allegorical references are rich with multiple meanings and are always open to interpretation. The allegorical verses of the Qur'an have been open to interpretation for the past fourteen centuries, and they will remain open to future generations. The Qur'an is a source of countless meanings due to such allegorical expressions; for this reason, the interpretation of the Qur'an in each century will give a sufficient explanation to people according to their level of understanding, and remain parallel to scientific discoveries, without contradictions arising.

THE FUTURE OF DARWINISM

We cannot think of the core of any superstitious claim as being totally empty and harmful. For if this were the case, a lot of people would not have pursued the most baseless schools of thoughts for so many years. The dumping ground of the history of thought is full of ideologies and philosophical movements that busied humanity with some crumbs of truth after which people chased for a time until they were abandoned, one by one. Some truths were distorted and misinterpreted, resulting in profound confusion and loss of belief among those who drifted, following after these movements. For instance, there was truth behind the notion of labor that Marxism put forward, but it was not everything. Meanwhile, as capitalism exalted capital, it stumbled into a different error by ignoring labor. For his part, Freud mistakenly attributed the thoughts of certain sick souls to all of humanity and credited the essence of the human being to the libido.

The fact that Darwinism, or in a broader sense, the evolutionary hypothesis, survived frictions with the church and became the dominant paradigm in a short time was mainly due to its stunning discovery of how some biological principles that exist among living beings operate. For example, it pointed out the existence of living beings as part of an integral whole, of a hierarchical system, and it drew attention to biological variation. Yet it could not provide the necessary explanations, and the interpretations which followed from it proceeded in a completely contrary direction.

Today, the doctrines of materialist and positivist philosophies have reached a bottleneck, and they cannot solve humanity's unease and global problems like terror. It is more frequently observed than at any other time that people are sincerely pursuing metaphysics, mystical beliefs, and religious thought. Many doctrines like Darwinism and its progeny "Social Darwinism" that serve in a way as an introduction to atheism have proved to lead humanity toward a dead end.

Special attention must be paid to prevent such metaphysical inclinations from assuming an anti-science character, which is as wrong as its opposite. We simply cannot ignore what the science of biology offers to us, nor can we allow it to be interpreted entirely within the evolutionist paradigm and thus abused as an instrument to promote atheism.

There are noteworthy efforts in the Christian as well as the Muslim worlds by means of establishing dialogue between religious thought and science. There is an increasing amount of sound research and a more constructive approach, as manifested in such publications as *Explore Evolution*[179] by organizations like the Templeton Foundation and Free Press. I would like to stress my conviction that a comprehensive interpretation of the Qur'anic verses concerning creation can reveal a perfect synthesis of religious thought and scientific research in a balance similar to the one Islam stipulates humans should observe between this world and the next. Approaching science and religion without separating them, as two sides of the same mirror, and regarding the cosmos in a holistic way, will make it possible to better understand the hierarchy in the creation, to benefit more from the horizons science will expand to, and to avoid erroneous thinking like "chance" that leads to atheism. I expect new developments in the Islamic world, and my hopes are supported extensively by the sincere research of God-believing scientists with common sense in the United States. Michael Behe, Michael Denton, Richard Milton, and Phillip Johnson are some of the authors who have produced notable works and have generated really significant breakthroughs in the West. Jeremy Rifkin is one of those authors too, and his book, *Algeny: A New Word, A New World,* points to the signs that reveal an increase in opposition to Darwinism. It will be more beneficial to refer to his words directly:

> Dr. Colin Patterson is a senior paleontologist at the British Natural History Museum, in London. Dr. Patterson is the author of the book, Evolution, and is recognized as the world's leading paleoichthyologist. On November 5, 1981, Dr. Patterson delivered a speech before a group of experts on evolutionary theory at the American Museum of Natural History. Dr. Patterson dared to suggest to his colleagues that the scientific theory that he and they had devoted a lifetime to was mere speculation, without any significant evidence to back it up. Here's how Dr. Patterson explained his change of mind concerning the theory of evolution: "Last year I had a sudden realization. For over twenty years I had thought I was working on evolution in some way. One morning

I woke up and something had happened in the night; and it struck me that I had been working on this stuff for twenty years and there was not one thing I knew about it. That's quite a shock, to learn that one can be so misled for so long. ... So for the last few weeks I've tried putting a simple question to various people and groups of people.... Can you tell me anything you know about evolution, any one thing, any one thing that is true? ... All I got ... was silence ... the absence of answers seems to suggest that ... evolution does not convey any knowledge, or, if so, I haven't yet heard it... I think many people in this room would acknowledge that during the last few years, if you had thought about it at all, you have experienced a shift from evolution as knowledge to evolution as faith. I know that it is true of me and I think it is true of a good many of you here.... Evolution not only conveys no knowledge but seems somehow to convey anti-knowledge."

Psychiatrist Karl Stern, of the University of Montreal, asks us all to detach ourselves from our preconceived biases and consider the merits of the Darwinian argument. The theory, says Stern, goes something like this: "At a certain point of time, the temperature of the Earth was such that it became most favorable for the aggregation of carbon atoms and oxygen with the nitrogen-hydrogen combination, and that from random occurrences of large clusters, molecules occurred which were most favorably structured for the coming about of life, and from that point, it went on through vast stretches of time until, through processes of natural selection, a being finally occurred which is capable of choosing love over hate, and justice over injustice, of writing poetry, like that of Dante, composing music, like that of Mozart, and making drawings, like those of Leonardo."

Stern's opinion of the evolutionary theory is not likely to win many friends within the scientific community. Speaking strictly from the point of view of a psychiatrist, he argues: "Such a view of cosmogenesis is crazy. And I do not at all mean "crazy" in the sense of slangy invective, but rather in the technical meaning of psychotic. Indeed, such a view has much in common with certain aspects of schizophrenic thinking."

Stern and Patterson are not alone. While biology teachers continue to teach the most up-to-date textbook version of Darwin's theory of evolution to the children of the 1980s, some of the high priests of biology have all but abandoned their own sacred texts.

Although unwilling to claim that evolution per se is a crazy idea, many of them are more than prepared to commit Darwin's version of it to the historical archives. Remarkably little has been written in the popular press about this rebellion in the making. The coup d'etat has unfolded rather quietly within the semi-sequestered domain of official academic conferences and scholarly journals. The first inkling that things were not well with Darwinism came, interestingly enough, during the centennial celebration of Darwin's theory, held at the University of Chicago in 1959. One of the speakers, paleontologist, Everett Claire Olson, of the University of California, let it be known that: "There exists as well, a generally silent group of students engaged in biological pursuits who tend to disagree with much of the current thought, but say and write little because they are not particularly interested, do not see that controversy over evolution is of any particular importance, or are so strongly in disagreement that it seems futile to undertake the monumental task of controverting the immense body of information and theory that exists in the formulation of modern thinking."

As to how many had actually deserted the ranks, Olson contended that it was difficult to judge the size and composition of this silent segment, but there is no doubt that the numbers are not inconsiderable. Overall, the present picture was that two hundred years of positivist and materialistic denial movements had found a new instrument for themselves to play around with.[180]

In effect, most scientists had begun to feel, in their minds and hearts, that evolution was a grand deception disguised as a "scientific" case. They would no longer comply with it willingly. Even the public began to enunciate a silent and giant "No!" in its stance and posture.

This is shown by the fact that most people in the world still turn towards religion, and people choose "cooperation" to solve the biggest global problems, although evolutionary theory has obviously been denying the Creator for decades; has produced a description of the universe as a dark, cold place, devoid of overriding control; and has encouraged humans to be enemies to each other, under the umbrella of "social Darwinism." In fact, the silence was shattered in 1959, and the dissenters then began to surface, one by one. Thus, once but a faint murmur, the opposition has now swollen into a chorus of discontent.

> Right now an intense struggle is going on within the profession, pitting the dyed-in-the-wool Darwinists against a new generation of theoreticians who are anxiously casting around for a more satisfactory explanation of the origin and development of species. The battle recently extended directly into London's Natural History Museum, long considered a bulwark of Darwinian thinking. At issue was a pamphlet published by the museum which qualified Darwinism by saying, "If the theory of Evolution is true. . ." "If" indeed! Much of the scientific community was aghast. To even suggest such a possibility – and coming from the British Natural History Museum – was enough to steam the bifocals of many a don at Cambridge, Oxford, Sussex, and other esteemed institutions throughout the kingdom. An editorial appearing in "Nature," the unofficial voice of the scientific establishment, rebuked museum officials in no uncertain terms. Noting that "most scientists would rather lose their right hand than begin a sentence with "if the theory of Evolution is true," the editorial asked rhetorically, "what purpose except confusion can be served by these weasel words?"

> Other establishment bastions have been caught up in the debate. For example, many years ago, G. A. Kerkut, a professor of Physiology and Biochemistry at the University of Southampton, England, published a book critical of Darwin's theory entitled, *Implications of Evolution*. Dr. Kerkut concluded: "The attempt to explain all living forms in terms of an evolution from a unique source, though a brave and valid attempt, is one that is premature and not satisfactorily supported by present-day evidence."

> An unusually candid review of the book, appearing in "*The American Scientist*," the official publication of the prestigious Sigma Xi scientific fraternity, acknowledged what many had long suspected but were afraid of to entertain, especially in print. Speaking to the book as well as to Darwin's theory, the review stated: "This is a book with a disturbing message; it points to some unseemingly cracks in the foundations. One is disturbed because what is said gives us the uneasy feeling that we knew it for a long time deep down but were never willing to admit this even to ourselves.... The particular truth is simply that we have no reliable evidence as to the evolutionary

sequence ... one can find qualified, professional arguments for
any group being the descendant of almost any other.... We have
all been telling our students for years not to accept any statement on its face value but to examine the evidence, and, therefore, it is rather a shock to discover we have failed to follow our
own sound advice."[181]

In fact, those who speak against Darwinism are sufficiently
populous to fill a book. Interestingly, some were against evolutionary theory from the beginning, and they confessed their conclusions
after a certain period of time, when the path was evidently viewed
as a dead end. Dr. Pierre P. Grassé, ex-president of the French
Academy of Sciences, and the editor of twenty-eight volumes of the
popular *Traité de Zoologie*, did not hesitate to attribute the designation of "pseudoscience" to evolution.[182] This declaration that evolutionary theory is a "pseudoscience" is now being heard with
increasing frequency. The British zoologist, Leonard Matthews,
expressed the concern of many of his colleagues in the introduction
to a 1971 edition of Darwin's book, *The Origin of Species* as follows:
"The fact of evolution is the backbone of biology, and biology is
thus in the peculiar position of being a science founded on an
unproved theory – is it then a science or faith?"[183] Writing in the
introduction to the 1956 publication of the same book by Darwin,
the entomologist, W. R. Thompson, reproached the "defenders of
the faith" for their unscientific conduct: "This situation, where men
rally to the defense of a doctrine they are unable to define scientifically, much less demonstrate with scientific rigor, attempting to
maintain its credit with the public by the suppression of criticism
and the elimination of difficulties, is abnormal and undesirable in
science."[184]

Another criticism came from Biology Professor, Edwin G.
Conklin, of Princeton University, who realized the pervasive sense
of religiosity that permeated the thinking of his colleagues: "The
concept of organic evolution is very highly prized by biologists, for

many of whom it is an object of genuinely religious devotion, because they regard it as a supreme integrative principle."[185]

Many scientists, encouraged by the lead of the "intelligent design movement," initiated especially by believing Christians in the US, are now able to express their ideas about evolution freely. The prominent individuals in this movement neither seem to represent any particular religious school of thought, nor do they appear necessarily to oppose secularity. They simply state that this universe has been designed by an intelligent agent. Yet it is not easy to entirely wipe out a taboo that has ossified and solidified in public mindset, so any start should be considered beneficial to preparing a peaceful atmosphere for discourse and study. It seems inevitable that many scientists will share this idea in the future. Thus, we can see that the means by which to withdraw evolutionary theory from the stage has already been set; for eventually, all will see that it is impossible for life to be explained by this theory, and they will surely abandon it. We may at least expect, in the near future, that evolutionary thought will become such a marginal movement that it will be left entirely aside. The process leading to such consequences has already begun, and the reason behind this is not solely the opposition of brave scientists. It is simply the case that the more we learn about life, the better we understand how complex it really is. Therefore, scientists are compelled to realize that that the countless intricate structures about which we learn more every day cannot be the products of purposeless, random mechanisms, as Darwin presumed.

Certainly, there is variation which emerges due to biological changes and which is refreshed by instant creations in the world of living beings. However, this variation does not happen in such a way as to allow transitioning from one species to another; rather, it occurs to increase the richness within a species, and thus, to exhibit the infinite power of God by providing thousands of reflections of His beautiful names. The genetic recombination mechanisms which cause diversity within species (subspecies and varieties) to occur, and biological principles such as natural selection and adaptation, do not prove evolution—rather, the truth is the other way around, for these all demonstrate the excellence of divine creation.

Natural selection, in fact, is a solution ordained by divine law to solve a sustainability problem, the food cycle or food pyramid, which is vital for the survival of living beings. In turn, adaptation mechanisms exhibit the potential for genetic change which was placed in the genetic program of living beings at the point of their creation and provided to ensure the continuation of the species under varying conditions.

As far as mutations are concerned, it must first be remembered that none of the useful changes in a living being's genomes occurs randomly; some of these mechanisms are provided to strengthen the immune system of the species; some serve to increase variation within the species (like meiosis and crossovers); and some are given to offer a veil of biological causes, which are appointed for living beings, such as ageing and death.

While scientists analyze the anatomical and physiological characteristics of organisms on one side, on the other side, they search for harmony between all these properties and look for ways in which those features serve the purpose of not just the species in question, but also of the population and the entire ecosystem. Nonetheless, the reason, heart and consciousness of a scientist will still incline him or her to behave "theologically," in a way, when interpreting data. This is because, even if modern science tries to separate philosophy from its study methods, a human being is a whole. Thus, sharing wisdom, or at least reflecting, is not just a necessity of being scientific, but also a most important aspect of being human (which is also vital for remaining alive). For this reason, scientists have to try to explain the divine wisdom of the organs, and the reason behind their shape, structure and functional features, and they have to see not only the design or plan in these structures and functions, but also the theology which can account for their optimization; in other words, a scientist has to see a particular creation as being favorable to its purpose.

Critically, a construction which is ideally suited to its purpose precludes origination by chance occurrence.

In musing on how a theory as scientifically bankrupt as Darwin's could ever have become the prevailing orthodoxy, Ludwig von Bertalanffy, who is acknowledged as one of the founders of the philosophy of biology, concluded as follows: "I think the fact that a theory so vague, so insufficiently verifiable and so far from the criteria otherwise applied in "hard" science, has become a dogma, can only be explained on sociological grounds. Society and science have been so steeped in the ideas of mechanism, utilitarianism and the economic concept of free competition, that instead of God, Selection was enthroned as ultimate reality"

Nowadays, it is almost impossible to find a place in popular science where evolutionary theory—which is a direct interest of the scientific world, and an indirect concern of the general public—is not being discussed. Both at the time that the theory was first put together, and throughout the process by which it has continually been reworked, it has certainly turned out to be something other than merely a biological idea. Everything that it touched was contaminated in the name of denying the Creator, by both direct and indirect efforts. Popular jokes about "how many scientists it takes to change a light bulb" are insufficient to describe the chaos and carnage which the contamination of this theory caused. Minds—and thus, hearts—were swayed into believing that the universe has no Creator, that it is ownerless. So, even though the universe is evidently a great work of art which follows from the Creator's unlimited knowledge, power, will and wisdom—as amply demonstrated through so many examples of intricate, orderly, and harmonious functions—humans have tried to solve the "mystery of creation" using only their own intelligence and accumulated knowledge, evaluating everything as though this marvelous, expressive universe had arisen by chance, and so betraying with their ingratitude their relationship with the Owner of the universe. Thus, the struggle to cover immeasurable truth with the veil of science has occurred, and continues. The response to those who are engaged in this struggle must be given within democratic means and in a tolerant way, not in the same way as evolutionists stand against creation.

It is impossible to accept a statement like, "It's just picking a quarrel, so let us not teach it," in regard to a matter which has been keeping the science world busy for 150 years, and which is still so widely discussed. If we said that ourselves, we would somehow be supporting a kind of ideological bigotry, or scientific dictatorship, and that would be wrong.

At this moment, however, the total reverse of such a situation is being experienced. In many cities and countries, evolution is taught exclusively in virtually all institutions, without giving an opportunity for those with dissenting views to articulate their opinions. Since evolutionary theory is discussed in all aspects, frequently crossing curriculum boundaries, faculty members at a number of universities, for instance, have had to endure the complaints of their colleagues, and the influence of a generally negative atmosphere in many countries, which has even made them the subject of academic investigations at times, such as has been the case in Turkey.

Recently, as a case in point, a professor of biology at a prominent university in Turkey was dismissed from his post simply because he held opinions opposed to evolutionary theory. A faculty member at another institution could not get his professorship for nine years due to his views about evolution; instead, his appointment was obstructed at two different universities because of the oppression that evolutionary ideology instilled in the academic community—even though his research was deemed sufficient for the position by the majority of members on the appointment committee. Unfortunately, the artificially tense environment which has developed in Turkey, in addition to pressure from the Turkish Higher Education Institution (YÖK), was applied to prevent him from securing from the Council of State what should have rightfully been his.

Other examples include situations where applications for associate professorships and professorships in biology, by candidates who question or even discuss evolution, have been obstructed with backstage activities and phone calls (and, if needed, fueled by rumors and exaggerations). Review committees have been warned not to make way for those who discuss or question evolution, while scientists have been oppressed openly or secretly due to the enforced dominance of evolutionary theory all around the world. In Germany, which is considered to be a democratic and highly developed state, excessive burdens can be imposed upon scientists. Prof. Dr. Wolf-Ekkehard Lönning, of the Max Planck Institute, was put under a ban after connecting his results, relating to aquatic plants, to the Creator, in his thousand-page report. In Turkey, some professors resort to using a pen name in order to be protected from academic harassment and intimidation when they write articles against Darwinism for popular magazines.

Voices which express the slightest reservation about evolution have been repeatedly threatened with academic obscurity, in reports by atheistic or materialistic stakeholders which raise complaints and accusations against them for being "religious fundamentalists," or "reactionaries"—even though some of those whose voices express such views are not in fact religious. I believe there are only a very small number of people in Turkey who lead this movement of oppression, in a completely militant atmosphere—and thus use the idea of evolution to cover up their own irreligious nature. However, since most of faculty members who are currently employed have been influenced by the stifling pressure of the atmosphere which has been established for many years, they do not dare to raise their voices if even they do not agree with evolutionary theory. Conversely, there are some colleagues who do believe in evolution, but still respect the rights of others to articulate their dissenting opinions.

Further, in Turkey, in spite of such a frustrating level of oppression, evolutionists, who see that the number of scientists who do not believe in evolution and are turning away from this theory is gradually increasing, have attempted to force the Ministry of Education

(MEB) by collecting signatures, specifically to augment the tension. Thus, oppressed by the Higher Education Institution, many academics cannot speak out and are being subjugated into silence. In fact, one faculty union made an attempt independently to collect signatures against evolutionists, but the faculty members of the biology department in particular apologized and held off due to the fear of incurring the Higher Education Institution's wrath and that they too could be forced to undergo the same difficult process that their colleagues were being subjected to.

Thus, in such an undemocratic institutional climate, evolutionists are able to do whatever they want without taking the rights of others into account. In addition to that, they have expended much effort on making the whole young generation atheistic in orientation, starting right from the level of elementary education, by trying to influence the Ministry of Education even more, though the subject of evolution is already given prominence in the course books of the Ministry of Education. However, such powerful evolutionary propaganda is still not sufficient for evolutionists, it seems, for they are militant in inclination and unyielding in Turkey, and there is doubt that such harshness exists in any other part of the world. Their goal is to drive evolution into all aspects of life as part of a total ideological program, as was once the case in the Soviet Union. They are not merely satisfied with their own disbelief, but rather, wish for everybody to disbelieve right along with them.

So, in the case of Turkey, my own country, which typifies the struggle against evolutionary dogma, what can be done? First, our universities have to become academically independent. Our scientists should be able to speak out about what they believe, and to believe in what they say. When the name of a course is "Evolution," it can be expected that evolution will be taught as if it were a definite law; so, first and foremost, the name of such courses should be changed. The most reasonable name for a course about subjects which cannot be scientifically tested, observed, or subjected to experimentation is "The Philosophy of Biology," which is offered at universities in many countries around the world. The faculty members teaching such a course should be neither obsessive nor fanatical. Rather, they should be democratic in orientation, tolerant, and respectful towards human rights.

Beside presenting findings which support evolution, instructors for such courses should also introduce totally contrary publications, or at least let the students bring such studies to class for the benefit of discussion. The faculty member could criticize whether a given publication agrees with scientific standards or not, but should not reprimand or stop students who bring articles which espouse views opposed to evolution, and which are routinely published in the most distinguished scientific magazines of the world.

The most powerful tool against evolutionary theory is probably the Internet. No matter where one is located in the world, all types of information—both favorable and unfavorable—can be had in just a few seconds. Therefore, the tense psychological atmosphere, created by the evolutionists' pressure tactics, has been diffused—and anyone who knows even a slight amount of a foreign language can step into the middle of debates relating to evolution which are happening all around the world, and thus become informed about many kinds of developments.

Also, if necessary, one lecturer could teach the course from evolutionary point of view, and later, another lecturer could teach the subject with reference to arguments which are contrary to evolution. Thus, one course would be taught from two different perspectives. If we do not consider students to be fools, then such a way of teaching would be very helpful, for students could listen to both teachers and come to their own determination on the matter by virtue of independent thinking.

Another important point is that debates should be performed frequently in open forums and panels, in a completely scientific fashion. In Turkey, we have unfortunately witnessed some rather tragic examples of poorly managed "scientific" debates. An atheist evolutionist announced during a television program, one day, that "Anyone who believes in God cannot be a scientist; such a person does not have a place in a university, and should be dismissed." How well scientific advances can still be accomplished in a country where such firm bigotry is experienced, and every subject is discussed in terms of ideologies, will determine the course by which our universities will grow, and thrive, in the future.

CLOSING REMARKS

We have seen that evolution—with all of its relevant arguments, deadlocks, and impossibilities—is nothing but a hypothesis. Though it is not possible to prove evolution, it has been enforced as an ideology and insisted upon in a desperate struggle by certain stakeholders to keep it alive. So, how can we come to a convincing conclusion concerning how the first living beings and the first human being were created without rejecting the information that the science of biology proposes? First of all, witnessing all the improbabilities which have been discussed, we have to admit that creation is a miracle. However, we can also say that even though it is a miracle, God, Who has infinite knowledge and power, used certain causes in the act of creation to veil His acts and operations. In addition, upon analyzing the verses of the Holy Qur'an related to creation taking "six days," and other similar verses about the duration and meaning of time, we may consider that our Lord first created the universe from nothing in phases within those days, and that only He knows the true duration of the process; and somewhere in that initial phase, we can say that He created the Milky Way somewhere in the universe, and then He created our solar system and our Earth at the most suitable places in the universe, to permit the most favorable conditions for life. We may say that in the consecutive "days," He created the atmosphere, the earth, mountains, seas, water and soil. After the Earth became favorable for life, He then created beings living in water, followed by living things on land, in a certain sequence. He created plants first, and then, He created herbivores, which would eat the plants, followed by carnivores, who would eat the herbivores. Finally, after the preparation of the Earth was completed, the first humans (omnivores), who could eat both plants and animals, were created.

In this way, we, too, can "order" creation according to a certain schema, but there could easily have been changes in the time span of each of these sequences, and the order of these sequences—and there could also have been many other events of which we cannot know. Since none of us witnessed the first creation, anything that can be said about this event and process cannot go beyond being simply an argument, or an alternative idea. Saying more pretentious things about this would amount to impertinence towards God. We can only make predictions that will not be contrary to our beliefs about this—and we must ensure that we neither contradict the essential data of science (like the fact that the Earth is round and rotates), nor articulate opinions which are disrespectful toward the Divine will.

In fact, the origins of life might have arisen as the result of a totally different creation process. It is even possible for the sequence of the creation process to be partially related to what evolutionists claim. Beyond this, however, the most important thing at the foundation of any understanding is that the conditions and materials of this world were used. If we call all of these things "causes" (i.e., the climate, soil, elements, heat, light, gravity, and so on), then we can successfully conclude that God effected the miracle of creation by making these causes veil His power at the precise time He commanded, through the outcomes of certain processes, in the specific amounts He ordered. While creation seems to have taken millions of years in our estimation—in terms of divine measure, everything might actually have occurred in a time span as brief as an instant. Yet however long that span was, and however we measure it, mindless and non-conscious causes could never have produced the act of creation on their own, to reach an agreement to form a living organism through their random efforts.

Whether the act of creation is a process (to us) or an instant (to God)—or even a phenomenon that can only ever be understood in a different dimension beyond our limited comprehension—it did happen through God's knowledge, power and will. In that we are familiar with the process of trial and error, we admit that it is impossible for idle changes, caused by natural forces, whose limits are not known, and by the movements of atoms, to convert one species to another, and thus to form a perfect new species by chance.

We believe in our Lord, Who has created the entire universe without any defects or faults, in the best and most excellent form, and Who made us—humankind—caliphs on this Earth, the most honorable of creatures, in order that we might be tested. We believe that He has thousands of names, and each of His names has 70,000 "degrees" (an allegory of multitude), and that those names, which can form infinite combinations, are manifested in an intricate fashion in every species. For instance, while the name, *Al-Razzaq* (Provider of food), appears in various degrees in plants, lions, mice and flies, the name *Al-Jamil* (He Who creates everything in beauty), is also manifested in different degrees in the very same living beings. Also, the name *Al-Hayy* (Lifegiver), is manifested in distinct levels in bacteria, viruses, plants, mushrooms, animals and humans. In addition, other names— such as *Al-Mudabbir* (He Who creates cautiously, trains and controls), *Al-Quddus* (He Who creates purely and perfectly, and keeps the universe clean), *Al-Musawwir* (He Who gives the appearance and form to His creatures, according to His will)—are also combined in different levels in each creature and cause everything, living or nonliving, to have various "degrees" of His names. In another example, His name, *Al-Sami'* (He Who hears everything), is recognized more prominently in an elephant, whale, mouse and shark, while His name, *Al-Basir* (He Who sees and watches everything), is manifested more in an eagle than in a rhinoceros; and so on.

The combination of thousands of beautiful names of God, in thousands of degrees, gives an opportunity for billions of different species to be created (for example, calculating it theoretically, to make it closer to reason, will yield $1,000^{70,000}$ possible names and degrees). In human beings, God's names are manifested in such a way as to make us suited to the highest position among all creatures. We cannot hear like a shark, nor can we see like an eagle—but the manifestations of all the names of God are characterized in our spiritual sensations and in our senses, which are unique to us, and in the innumerable bounties which any one of us may recognize and contemplate for ourselves—namely the human mind, reason, heart, perception, and intuition.

NOTES

1   Jeremy Rifkin, *Algeny: A New Word, A New World*. (Penguin: 1984).

2   Julian Huxley, "At Random – A Television Preview," *Evolution After Darwin*, (University of Chicago Press 1960) ed. Sol Tax, Vol. I, p. 42.

3   Karl Raimund Popper, *Unended Quest: An Intellectual Autobiography*. (Illinois: Open Court, 1976) The Library of Living Philosophers, Vol. 1, p. 133.

4   Phillippe Janvier, "Phylogenetic classifications of living and fossil vertebrates." *Bulletin de la Societe Zoologique de France*, 1997, Vol. 122, pp. 341–354.

5   Aleksandr Ivanovich Oparin, *Life: Its Nature, Origin and Development*. (London: Oliver&Boyd 1961), p. 33. Translated from Russian by Ann Synge.

6   Francis Darwin (ed.), "Letter to Asa Gray." *The Life and Letters of Charles Darwin*, (New York: Appleton, 1887), Vol. II, p. 67.

7   Joel Achenbach, "Life beyond Earth." *National Geographic* 2000, January, Washington.

8   Bediüzzaman Said Nursi, (1976). *Lem'alar*. Risale-i Nur Külliyatından, (Istanbul: Sözler Yayınevi, 1976) pp.166–183.

9   Herbert Spencer, *First Principles of a New System of Philosophy* (New York, Appleton, 1872). Two volumes.

10 Gertrude Himmelfarb, *Darwin and the Darwinian Revolution* (New York: W. W. Norton & Company, 1959.)

11 Norman Macbeth, *Darwin Retried: An Appeal to Reason* (Boston: Gambit, 1971), pp. 99–100.

12 R. L. Wysong, *The Creation-Evolution Controversy*. (East Lansing, MI: Inquiry Press, 1976), p. 422.

13 Francis Darwin (ed.), "Letter to Asa Gray." *The Life and Letters of Charles Darwin* (London: John Murray, 1888), Vol. 2, p. 273.

14 Lester J. McCann, *Blowing the Whistle on Darwinism*, 1986, p. 49.

15 Himmelfarb 1959.

16 Conrad Hal Waddington, *The Strategy of the Genes* (London: Allen-Unwin, 1957), pp. 64–65.

17 Pierre-Paul Grassé, *Evolution of Living Organisms* (New York: Academic Press, 1977), p. 202.

18 Rifkin 1984.

19 John Arthur Thomson, Patrick Geddes, *Life: Outlines of General Biology* (London: Williams & Norgate 1931), Vol. II, p. 1317.

20 Michael Denton, *Evolution: A Theory in Crisis*. (London: Burnett Books, 1985).

21 ibid.

22 ibid.

23 ibid.

24 Richard Milton, *Shattering the Myths of Darwinism* (Vermont: Park Street Press, 1997). Francis Hitching, *The Neck of the Giraffe: Where Darwin Went Wrong* (New York: Ticknor and Fields, 1982), p. 204.

25 Luke Harding, "History of modern man unravels as German scholar is exposed as fraud," *The Guardian*, February 19, 2005.

26 Matthias Schulz, "Die Regeln Mache Ich," *Der Spiegel*, August 16, 2004.

27 "On Campus," Alleged skullduggery, Random Samples. *Science*, Vol 305, Issue 5688, p. 1237, August 27, 2004.

28 Tony Paterson, "Neanderthal Man never walked in northern Europe." www.telegraph.co.uk/news/main.jhtml?xml=/news/2004/08/22/wnean22.xml. August 22, 2004.

29 Harding 2005.

30 Milton 1997; Hitching 1982.

31 Richard E. Lingenfelter, "Production of C-14 by Cosmic 8 Ray Neutrons." *Reviews of Geophysics*, 1:51, February, 1963.

32 Hans E. Suess, "Secular Variations in the Cosmic-Ray produced Carbon-14 in the Atmosphere and Their Interpretations." *Journal of Geophysical Research*, 70:5947, December 1, 1965.

33 V. R. Switzer, "Radioactive Dating and Low-level Counting," *Science*, 157:726, August 11, 1967.

34 Melvin A. Cook, "Where is the Earth's Radiogenic Helium," *Nature*, 179:213, January 26, 1957.

35 Frank Hole and Robert Heizer, *Prehistoric Archaeology: A Brief Introduction*. Harcourt College Publishers, 3rd ed. 1977.

36 R. W. Fairbridge, "Holocene." In *Encyclopaedia Britannica*, 1984.

37 Jacques Evin, "Le temps et la chronométrie en archéologie." *Histoire et Mesure*. Vol. IX - N° 3/4, Archéologie II, 1994.

38 Peter Ward and Donald Brownlee, *Rare Earth* (New York: Copernicus, 2000).

39 Stephen Jay Gould, *Wonderful Life: The Burgess Shale and the Nature of History*, (New York: W. W. Norton & Company, 1989).

40 Denton 1985.

41 Éric Buffetaut, *Grandes Extinctions et Crises Biologiques* (Milan: Mentha, 1992), p. 53.

42 ibid.

43 J. J. Jaeger, "Les Catastrophes Géologiques," in *La Mémoire de la Terre*, (Seuil, 1992), pp. 139–148.

44 ibid.

45 Douglas Erwin, "The Mother of Mass Extinctions." *Scientific American*. July 1996, pp. 56–62.

46 Vincent Courtillot, "Une éruption volcanique?" *Dossiers pour la Science*, Hors Série, Septembre–Novembre, 1990, pp. 84–92.

47 Charles. B. Officer and Charles L. Drake, "The Cretaceous-Tertiary Transition," *Science* 1983, 219: 1383–1390.

48 Louis de Bonis, *Evolution et extinction dans le règne animal*, (Paris: Masson, 1991).

49 David Raup and Jack Sepkoski, "Periodicity of Extinctions in the Geologic Past." *Proceedings of the National Academy of Science*, 1984, 81: 801–805.

50 Steven M. Stanley, "Mass Extinctions in the Ocean." *Scientific American*, No: 6 (June 1984), pp. 64–72.

51 Buffetaut 1992.

52 Thomas Kuhn, *The Structure of Scientific Revolutions* (Chicago: University of Chicago Press, 1962).

53 B. J. Stahl, *Vertebrata History. Problems in Evolution*. (New York: McGraw-Hill, 1985), p. 146.

54 Charles Darwin, *The Origin of Species*, Modern Library Paperback Edition, 1993. p. 167; Random House, Inc. 1998, USA.

55 John Sepkoski, Jr., "Rates of speciation in the fossil record." *Philosophical Transactions of the Royal Society of London B: Biological Sciences*, 353 (1366). 315–326.

56 David Raup, "Conflicts between Darwin and Paleontology," *Field Museum of Natural History Bulletin*, vol. 50. No. 1, 1979, pp. 22–29.

57 George Gaylord Simpson, *The Major Features of Evolution*. (New York: Columbia University Press, 1961) pp. 359–360.

58 W. E. Swinton, "The Origin of Birds" in *Biology and Comparative Physiology of Birds*, A. J. Marshall (edited by) (New York: Academic Press, vol. 1, p. 1–14.

59 Bone Bonanza, "Early Bird and Mastodon," *Science News*, 112 (September 2, 1977), p. 198.

60 M. S. Germain, "Qui est l'ancêtre des oiseaux?" *Science et Vie*, Paris: 1999, No: 977.

61 "Old Bird," *Discover*, (March 1997), p. 21.

62 A. Feduccia, L. Martin, Z. Zhou, and L. Hou, "Birds of a Feather," *Scientific American* (June 1998), 8.

63 D. B. Kitts, "Paleontology and Evolution Reconsidered," *Paleobiology*, 1977, 3, p. 115.

64 H. J. Jerison, *Evolution of the Brain and Intelligence* (New York and London: Academic Press, 1973).

65 R. Levin, "Bones of Mammals, Ancestors Fleshed Out." *Science*, Vol. 212, 26 June 1981, p. 1492.

66 George Gaylord Simpson, *Life Before Man*, (New York: Time-Life Books, 1972), p. 42.

67 Mark Ridley, "Who doubts evolution?" *New Scientist*, 25 June 1981, Vol. 90, pp. 830–832.

68 "Did Darwin Get it Wrong?" PBS Television Show, November 1, 1981. WGBH Transcripts, 125.

69 Boyce Rensberger, *Houston Chronicle,* November 5, 1980, Part 4, p. 15.

70 C. Patterson, *Harper's*, February 1984. p. 60.

71 G. R. Taylor, *The Great Evolution Mystery* (New York: Harper & Row, 1983), p. 230.

72 George Gaylord Simpson, *Horses*, (Oxford University Press, 1961).

73 Raup 1979.

74 Leonard Huxley, *Life and Letters of Thomas Henry Huxley* (London: MacMillan, 1900).

75 Karl Raimund Popper, "Darwinism as a metaphysical research programme" *Methodology and Science*, 1976, 9, 103–119.

76 Lambert Beverly Halstead, "Museum of Errors," *Nature*, November 20, 1980, p. 208.

77 S. J. Gould, N. Eldredge, "Punctuated Equilibra: The Tempo and the Mode of Evolution Reconsidered," *Paleobiology*, 1977, 3: p. 115–151.

78 "Missing, Believed Nonexistent," The Guardian Weekly, November 26, 1978, vol. 119, no 22, p.1, in Denton, 1988.

79 Gabriel Dover, "Molecular drive: a cohesive mode of species evolution," *Nature*, 1982, 229, 111–117.

80 W. S. Weiner, K. P. Oakley, W. E. Le Gros Clark, "The Solution of the Piltdown Problem," *Bulletin of the British Museum (Natural History) Geology Series*, 1953, Vol. 2, No. 3.

81 William K. Gregory, "Hesperopithecus Apparently Not an Ape nor A Man," *Science*, 1927, Vol. 66, December, p. 579.

82 Theodosius Dobzhansky, *Mankind Evolving. The Evolution of the Human Species*, (New Haven and London: Yale University Press, 1969).

83 B. Wood and A. Brooks, "We are what we ate," *Nature*, 1999, Vol. 400, no: 6741, 15 July 1999.

84 Michael J. Behe, *Darwin's Black Box: The Biochemical Challenge to Evolution*, Free Press, 1996, p. 307.

85 H. Watanabe and E. Fujiyama, et. al. "DNA sequence and comparative analysis of chimpanzee chromosome 22," *Nature*, 2004, 429, 382–388.

86 Laura Nelson, "First chimp chromosome creates puzzles," *Nature Science Update*, May 27, 2004.

87 Jean Chaline, "L'Evolution Biologique Humaine," *Que Sais-Je?*, (Paris: Presses Universitaires de France, 1982).

88 P. Darlu, "A quelle distance sommes-nous de nos voisins singes?" *Science & Vie*, Hors Série, Trimestriel, no. 200, Septembre. Paris, 1997.

89 Jean Staune, "L'évolution condamne Darwin." Excerpt from the interview with *Jean Dorst. Figaro Magazine*. October 26, 1991, p. 15.

90 Gretchen Vogel, "Objection 2: Why Sequence the Junk?" *Science*, February 16, 2001.

91 H. Renauld, S. M. Gasser, "Heterochromatin: a meiotic matchmaker," *Trends in Cell Biology 7* May 1997, pp. 201–205.

92 E. Zuckerkandl, "Neutral and Nonneutral Mutations: The Creative Mix-Evolution of Complexity in Gene Interaction Systems," *Journal of Molecular Evolution*, 1997, 44, p. 2–8.

93 Elizabeth Pennisi, *Science News*, December 10, 1994.

94 M. J. Beaton and T. Cavalier-Smith, "Eukaryotic non-coding DNA is functional: evidence from the differential scaling of cryptomonal genomes," *Proc. R. Soc. Lond. B.* 1999, 266: 2053–2059.

95 L. L. Sandell and V. A. Zakian, "Loss of a yeast telomere: arrest, recovery, and chromosome loss" *Cell* 1993, 75 (4) 729–739.

96 S. J. Ting, "A binary model of repetitive DNA sequence in Caenorhabditis elegans." *DNA Cell Biol.* 1995, 14: 83–85.

97 E. R. Vandendries, D. Johnson, R. Reinke, "Orthodenticle is required for photoreceptor cell development in the Drosophila eye." *Dev Biol* 1996, 173: 243–255.

[98] J. Kohler, S. Schafer-Preuss, D. Buttgereit, "Related enhancers in the intron of the beta1 tubulin gene of Drosophila melanogaster are essential for maternal and CNS-specific expression during embryogenesis." *Nucleic Acids Res* 1996, 24: 2543–2550.

[99] S. Hirotsune, N. Yoshida, A. Chen, L. Garrett, F. Sugiyama, S. Taka-hashi, K. Yagami, A. Wynshaw-Boris, A. Yoshiki, "An expressed pseudo-gene regulates the messenger-RNA stability of its homologous coding gene." *Nature* 2003, 423: 91–96.

[100] W. Makalowski, "Not Junk After All" *Science*, 23 May 2003, Vol. 300. no. 5623, pp. 1246–1247.

[101] S. Fisher, E. A. Grice, M. Ryan, R. M. Vinton, L. Seneca, S. L. Bessling, S. Andrew, A. S. Mccallion, "Conservation of RET Regulatory Function from Human to Zebrafish Without Sequence Similarity" *Science Express* March 23, 2006 (Online). This work first appeared in the press as "Junk DNA may not be so junky after all."

[102] Rifkin 1984.

[103] Keith Stewart Thomson, "Ontogeny and Phylogeny Recapitulated," *American Scientist*, Vol. 776, May–June 1988, p. 273.

[104] Hannington Enoch, *Evolution or Creation*, (London: Evangelical Press, 1968), pp. 57–58.

[105] Thomas Stanley Westoll. Proceedings from the British Association Meeting at Edinburgh, August 10, 1951.

[106] W. J. Bock, "Evolution by Orderly Law," *Science*, Vol. 164, May 9, 1969, pp. 684–685.

[107] Gavin Rylands de Beer, *Embryos and Ancestors* (New York: Oxford University Press, 1954).

[108] R. Danson, "Evolution" *New Scientist*, 1971, No. 49.

[109] Jonathan Wells, *Icons of Evolution: Science or Myth? Why Much of What We Teach about Evolution Is Wrong* (Washington DC: Regnery Press, 2000).

[110] George Gaylord Simpson, W. Beck, *An Introduction to Biology*, Harcourt Brace and World, New York, 1965, p. 241.

[111] Ken McNamara, "Embryos and Evolution," *New Scientist*, October 16, 1999.

[112] Michael K. Richardson et al., "Haeckel, Embryos, and Evolution," *Science*, May 15, 1998 280:983–985.

[113] Michael K. Richardson, J. Hanken, M. L. Gooneratne et al., "There is no highly conserved embryonic stage in the vertebrates, implications for current theories of evolution and development," *Anatomy and Embryology*, 1997, 196, 91–106.

[114] Michael K. Richardson, "Haeckel's Embryos, Continued," *Science*, 1998, 281, 1289.

[115] Michael K. Richardson and Gerhard Keuck, "A question of intent: when is a 'schematic' illustration a fraud?" *Nature* 2001, 410:144.

[116] Michael K. Richardson and Gerhard Keuck, "Haeckel's ABC of evolution and development," *Biological Reviews of the Cambridge Philosophical Society* 2002, 77, pp. 495–528.

[117] Wilhelm His, *Die Anatomie menschlicher Embryonen*, (Leipzig: Vogel, 1880).

[118] Ludwig Rutimeyer, "Rezension zu Haeckel, Ernst, Naturliche Schöpfungsgeschichte," (Berlin: 1868), *Archiv für Anthropologie* 3, 301–302.

[119] Richardson and Keuck 2002.

[120] Elizabeth Pennisi, "Haeckel's Embryos: Fraud Rediscovered," *Science* Vol. 277, No. 5331, p. 1435, September 5, 1997.

[121] J. M. Oppenheimer, "Haeckel's variations on Darwin," *Biological Metaphor and Cladistic Classification: An Interdisciplinary Perspective*, p. 123–135. Edited by H. M. Hoenigswald and L. F. Wiener (Philadelphia: University of Pennsylvania Press, 1987).

[122] Hitching 1982.

[123] Wysong 1976.

[124] Arslan Mayda, "İşe yaramaz zannedilen kuyruk sokumu." *Sızıntı*, 1997, No. 227, Izmir.

[125] Jerry Bergman and George Howe, *Vestigial Organs are Fully Functional* (Terre Haute: Creation Research Society Books, 1990).

[126] ibid.

[127] S. Maeda and G. Mogi, "Functional Morphology of Tonsillar Crypts in Recurrent Tonsillitis," *Acta Otolaryngo (Stockh) Suppl*, 1984, 416:7–19.

[128] Bergman and Howe 1990.

[129] Denton 1985.

[130] Frank B. Salisbury, "Natural Selection and the Complexity of the Gene," *Nature*, 1969, 224: 342.

[131] Georges Lemaitre, "Un univers homogène de masse constante et de rayon croissant, rendant compte de la vitesse radiale des nébuleuses extragalactiques," *Annales de la Société scientifique de Bruxelles* 1927, 47: 49–59.

[132] Alexander Friedman, "Über die Krümmung des Raumes," *Zeitschrift für Physik* 1922, 10: 377–86.

[133] Edwin Hubble, "A Relation between Distance and Radial Velocity among Extra-galactic Nebulae," *Proceedings of the National Academy of Sciences* 1929, 15: 168–73.

134 Arthur Eddington, *The Expanding Universe*, (New York: Macmillan, 1933), p. 124.

135 Fred Hoyle, *Frontiers in Astronomy*, (London: William Heinemann Ltd, 1955).

136 A. A., Penzias and R. W. Wilson, "A Measurement of Excess Antenna Temperature at 4080 Mc/s," *Astrophysics Journal* 1965, 142, p. 419.

137 J. M. Caron, A. Gauthier, A. Schaaf, J. Ulysse, and J. Wozniak, *La Planète Terre* (Paris: Editions Ophrys, 1992), p. 271.

138 Oparin 1961.

139 Stanley L. Miller, "A Production of Amino Acids Under Possible Primitive Earth Conditions," *Science*, Vol. 117, May 15, 1953. No: 3046, p. 528–529.

140 Stanley L. Miller and H. C. Urey, "Organic Compound Synthesis on the Primitive Earth," *Science*, 1959, 130, 245.

141 Rifkin 1984.

142 R. B. Bliss and G. E. Parker, *Origin of Life: Evolution–Creation*, (California: Creation Life Publishers, 1979), p. 14.

143 P. T. Mora, "The Folly of Probability" in *The Origins of Prebiological Systems and Their Molecular Matrices*, edited by Sidney W. Fox (New York: Academic Press, 1965), p. 41.

144 Marcel Florkin, "Ideas and Experiments in the Field of Prebiological Chemical Evolution," *Comprehensive Biochemistry*, 1975, 29B, 231–260, pp. 241–242.

145 Heinrich D. Holland, "Model for the Evolution of the Earth's Atmosphere" in *Petrologic Studies: A Volume in Honor of A. F. Buddington*, edited by A. E. J. Engel, Harold L. James and B. F. Leonard, (New York: Geological Society of America, 1962), pp. 448–449.

146 Philip H. Abelson, "Chemical Events on the Primitive Earth." *Proceedings of National Academy of Science*, 1966, Vol. 55, pp. 1365–1372.

147 Sidney W. Fox and Klause Dose, *Molecular Evolution and the Origin of Life*, Revised Edition, (New York: Marcel Dekker, 1977), pp. 43, 74–76.

148 Heinrich D. Holland, "When did the Earth's atmosphere become oxic? A Reply," *Geochemical News*, 1999, 100, pp. 20–22.

149 R. T. Brinkman, "Dissociation of Water Vapor and Evolution of Oxygen in the Terrestrial Atmosphere," *Journal of Geophysical Research*, 1969, Vol. 74: 23, pp. 53–66.

150 Harry Clemmey and Nick Badham, "Oxygen in the Precambrian Atmosphere: An evaluation of the geological evidence," *Geology* 1982, 10, pp. 141–146.

[151] Wysong 1976.

[152] Abelson 1966.

[153] J. P. Ferris and D. E. Nicodem, "Ammonia Photolysis and the Role of Ammonia in Chemical Revaluation," *Nature*, 1972, Vol. 238, p. 269.

[154] J. P. Ferris and C. T. Chen, "Photochemistry of Methane, Nitrogen and Water Mixture as a Model for the Atmosphere of the Primitive Earth," *Journal of American Chemical Society*, 1975, Vol. 97:11, pp. 2962–2967.

[155] Stanley L. Miller, "Current Status of the Prebiotic Synthesis of Small Molecules," *Molecular Evolution of Life*, 1986, p. 7.

[156] ibid.

[157] A. Katchalsky, "Prebiotic synthesis of biopolymers on inorganic templates," *Naturwiss*, 1973, 60: 215–220.

[158] T. E. McMullen, "Problems with chemical origins of life theories," Excerpts from his lectures between April 16, 1993 and April 3, 1995 at South Carolina University. (http://www2.gasou.edu/facstaff/etmcmull/CHEM.htm).

[159] A. E. Wilder-Smith, *The Natural Sciences Know Nothing of Evolution*, (California: Master Books, 1981), pp. 9–89.

[160] G. A. Kerkut, *The Implications of Evolution*, (London: Pergamon Press, 1960).

[161] Simpson, *The Major Features of Evolution*, 1961.

[162] ibid.

[163] McMullen 1998.

[164] Fred Hoyle and Chandra Wickramasinghe, *Evolution from Space*, (London: J. M. Dent and Sons, 1981), p. 24.

[165] Fred Hoyle. *The Intelligent Universe*, (London: Michael Joseph Ltd, 1982), 256 pp.

[166] Francis Crick, *Life Itself: Its Origin and Nature*, (New York: W.W. Norton, 1982), 192 pp.

[167] V. H. Mottran, "In the Organ Corporation," *Liner*, April 22, 1948.

[168] Pierre Lecompte du Nouy, *Human Destiny*, (London: Longmans Gren and Co., 1947), First Ed. pp. 33–34.

[169] Harold J. Morowitz, *Energy Flow in Biology*, (New York: Academic Press, 1968), p. 179.

[170] A. G. Cairns-Smith, *The Life Puzzle*, (Edinburgh: Oliver and Boyd, 1971).

[171] George Gamow, *The Creation of the Universe*, revised edition, (New York: Viking 1961).

[172] James F. Coppedge, *Evolution: Possible or Impossible?* (Northridge, California: Probability Research in Molecular Biology, 1993), 107, 114, 115.

[173] ibid.

[174] Salisbury 1969

[175] Hubert P. Yockey, "A Calculation of the Probability of Spontaneous Biogenesis by Information Theory," *Journal of Theoretical Biology*, 1977, 67: 377–398. This work has later developed into a book: *Information Theory and Molecular Biology*, (Cambridge University Press, 1992), p. 408.

[176] Carl Sagan, "Life," *Encyclopedia Britannica*. (New York: Encyclopedia Britannica, 1997), 22: 967.

[177] George Gaylord Simpson, *The Meaning of Evolution*. Revised Edition. (New Haven, Connecticut: Yale University Press, 1967).

[178] Rifkin 1984.

[179] Stephen C. Meyer, Scott Minnich, Jonathan Moneymaker, Paul A. Nelson, and Ralph Seelke, *Explore Evolution: The Arguments for and Against Neo-Darwinism*, (Melbourne: Hill House Publishers, c/o O'Brien & Partners, 2007).

[180] Rifkin 1984.

[181] ibid.

[182] Grassé 1977.

[183] L. Harrison Matthews, from the "Introduction" to *The Origin of Species* by Charles Darwin, 1971 edition. (London: J. M. Dent and Sons, 1971), p. xi.

[184] W. R. Thompson, from the "Introduction" to *The Origin of Species* by Charles Darwin, 1956 edition. (New York: E. P. Dutton, 1956).

[185] Edwin Grant Conklin, *Man Real and Ideal*, (New York: Scripner's, 1943), p. 147.

BIBLIOGRAPHY

Abelson, Philip H. "Chemical Events on the Primitive Earth." *Proceedings of National Academy of Science*, 1966, Vol. 55.

Achenbach, Joel. "Life beyond Earth." *National Geographic* January 2000, Washington.

Beaton M. J., and T. Cavalier-Smith. "Eukaryotic non-coding DNA is functional: evidence from the differential scaling of cryptomonal genomes," *Proc. R. Soc. Lond. B.* 1999, 266.

Beer, Gavin Rylands de. *Embryos and Ancestors*. New York: Oxford University Press, 1954.

Behe, Michael J. *Darwin's Black Box: The Biochemical Challenge to Evolution*, Free Press, 1996.

Bergman, Jerry, and George Howe, *Vestigial Organs are Fully Functional*. Terre Haute: Creation Research Society Books, 1990.

Bliss, R. B., and G. E. Parker, *Origin of Life: Evolution–Creation*. California: Creation Life Publishers, 1979.

Bock, W. J. "Evolution by Orderly Law," *Science*, Vol. 164, May 9, 1969.

Bonanza, Bone. "Early Bird and Mastodon." *Science News*, 112. September 2, 1977.

Brinkman, R. T. "Dissociation of Water Vapor and Evolution of Oxygen in the Terrestrial Atmosphere," *Journal of Geophysical Research*, 1969, Vol. 74: 23.

Bonis, Louis de. *Evolution et extinction dans le règne animal*. Paris: Masson, 1991.

Buffetaut, Éric. *Grandes Extinctions et Crises Biologiques*. Milan: Mentha, 1992.

Cairns-Smith, A. G. *The Life Puzzle*. Edinburgh: Oliver and Boyd, 1971.

Caron, J. M., A. Gauthier, A. Schaaf, J. Ulysse, and J. Wozniak, *La Planète Terre*. Paris: Editions Ophrys, 1992.

Chaline, Jean. "L'Evolution Biologique Humaine." *Que Sais-Je?* Paris: Presses Universitaires de France, 1982.

Clemmey, Harry, and Nick Badham. "Oxygen in the Precambrian Atmosphere: An evaluation of the geological evidence," *Geology* 1982, 10.

Conklin, Edwin Grant. *Man Real and Ideal*. New York: Scripner's, 1943.

Cook, Melvin A. "Where is the Earth's Radiogenic Helium," *Nature*, 179:213, January 26, 1957.

Coppedge, James F. *Evolution: Possible or Impossible?* Northridge, California: Probability Research in Molecular Biology, 1993.

Courtillot, Vincent. "Une éruption volcanique?" *Dossiers pour la Science*, Hors Série, Septembre–Novembre, 1990.

Crick, Francis. *Life Itself: Its Origin and Nature*, New York: W.W. Norton, 1982.

Danson, R. "Evolution" *New Scientist*, 1971, No. 49.

Darlu, P. "A quelle distance sommes-nous de nos voisins singes?" *Science & Vie*, Hors Série, Trimestriel, no. 200, September. Paris, 1997.

Darwin, Charles. *The Origin of Species*, Modern Library Paperback Edition, 1993.

———. *The Origin of Species*, Random House, Inc. 1998, USA.

Darwin, Francis (ed.). "Letter to Asa Gray." *The Life and Letters of Charles Darwin*. New York: Appleton, 1887. Vol. II.

———. "Letter to Asa Gray." *The Life and Letters of Charles Darwin*. London: John Murray, 1888. Vol. 2, p. 273.

Denton, Michael. *Evolution: A Theory in Crisis*. London: Burnett Books, 1985.

"Did Darwin Get it Wrong?" PBS Television Show, November 1, 1981. WGBH Transcripts, 125.

Dobzhansky, Theodosius. *Mankind Evolving. The Evolution of the Human Species*, New Haven and London: Yale University Press, 1969.

Dover, Gabriel. "Molecular drive: a cohesive mode of species evolution." *Nature*, 1982, 229.

Eddington, Arthur. *The Expanding Universe*, New York: Macmillan, 1933.

Enoch, Hannington. *Evolution or Creation*, London: Evangelical Press, 1968.

Erwin, Douglas. "The Mother of Mass Extinctions." *Scientific American*. July 1996.

Evin, Jacques. "Le temps et la chronométrie en archéologie." *Histoire et Mesure*. Vol. IX - N° 3/4, Archéologie II, 1994.

Fairbridge, R. W. "Holocene." In *Encyclopaedia Britannica*, 1984.

Feduccia, A., L. Martin, Z. Zhou, and L. Hou. "Birds of a Feather." *Scientific American*. June 1998.

Ferris, J. P., and D. E. Nicodem. "Ammonia Photolysis and the Role of Ammonia in Chemical Revaluation," *Nature*, 1972, Vol. 238.

Ferris, J. P., and C. T. Chen. "Photochemistry of Methane, Nitrogen and Water Mixture as a Model for the Atmosphere of the Primitive Earth," *Journal of American Chemical Society*, 1975, Vol. 97:11.

Fisher, S., E. A. Grice, M. Ryan, R. M. Vinton, L. Seneca, S. L. Bessling, S. Andrew, A. S. Mccallion, "Conservation of RET Regulatory Function from Human to Zebrafish Without Sequence Similarity" *Science Express* March 23, 2006 (Online). This work first appeared in the press as "Junk DNA may not be so junky after all."

Florkin, Marcel. "Ideas and Experiments in the Field of Prebiological Chemical Evolution," *Comprehensive Biochemistry*, 1975, 29B, 231–260.

Fox, Sidney W., and Klause Dose. *Molecular Evolution and the Origin of Life*, Revised Edition, New York: Marcel Dekker, 1977.

Friedman, Alexander. "Über die Krümmung des Raumes," *Zeitschrift für Physik* 1922, 10.

Gamow, George. *The Creation of the Universe*, revised edition. New York: Viking, 1961.

Germain, M. S. "Qui est l'ancêtre des oiseaux?" *Science et Vie*, 1999, No: 977. Paris: 1999, No: 977.

Gould, Stephen Jay. *Wonderful Life: The Burgess Shale and the Nature of History*, New York: W. W. Norton & Company, 1989.

Gould, Stephen Jay, and N. Eldredge, "Punctuated Equilibra: The Tempo and the Mode of Evolution Reconsidered," *Paleobiology*, 1977, 3.

Grassé, Pierre-Paul. *Evolution of Living Organisms*. New York: Academic Press, 1977.

Gregory, William K. "Hesperopithecus Apparently Not an Ape nor A Man," *Science*, 1927, Vol. 66, December.

Halstead, Lambert Beverly. "Museum of Errors," *Nature*, November 20, 1980.

Harding, Luke. "History of modern man unravels as German scholar is exposed as fraud," *The Guardian*, February 19, 2005.

Hirotsune, S., N. Yoshida, A. Chen, L. Garrett, F. Sugiyama, S. Takahashi, K. Yagami, A. Wynshaw-Boris, A. Yoshiki, "An expressed pseudogene regulates the messenger-RNA stability of its homologous coding gene." *Nature* 2003, 423.

His, Wilhelm. *Die Anatomie menschlicher Embryonen*, Leipzig: Vogel, 1880.

Holland, Heinrich D. "Model for the Evolution of the Earth's Atmosphere" in *Petrologic Studies: A Volume in Honor of A. F. Buddington*. Edited by A. E. J. Engel, Harold L. James and B. F. Leonard. New York: Geological Society of America, 1962.

———. "When did the Earth's atmosphere become oxic? A Reply," *Geochemical News*, 1999, 100.

Himmelfarb, Gertrude. *Darwin and the Darwinian Revolution*. New York: W. W. Norton & Company, 1959.

Hitching, Francis. *The Neck of the Giraffe: Where Darwin Went Wrong*. New York: Ticknor and Fields, 1982.

Hole, Frank, and Heizer, Robert. *Prehistoric Archaeology: A Brief Introduction.* Harcourt College Publishers, 1977, 3rd ed.

Hoyle, Fred. *Frontiers in Astronomy,* London: William Heinemann Ltd, 1955.

Hoyle, Fred, and Chandra Wickramasinghe, *Evolution from Space.* London: J. M. Dent and Sons, 1981.

Hoyle, Fred. *The Intelligent Universe,* London: Michael Joseph Ltd, 1982.

Hubble, Edwin. "A Relation between Distance and Radial Velocity among Extra–galactic Nebulae," *Proceedings of the National Academy of Sciences* 1929, 15.

Huxley, Julian. "At Random – A Television Preview," *Evolution After Darwin.* University of Chicago Press 1960. Edited by Sol Tax, Vol. I.

Huxley, Leonard. *Life and Letters of Thomas Henry Huxley.* London: MacMillan, 1900.

Jaeger, J. J. "Les Catastrophes Géologiques," in *La Mémoire de la Terre.* Seuil, 1992.

Janvier, Phillippe. "Phylogenetic classifications of living and fossil vertebrates." *Bulletin de la Societe Zoologique de France,* 1997, Vol. 122.

Jerison, H. J. *Evolution of the Brain and Intelligence.* New York and London: Academic Press, 1973).

Katchalsky, A. "Prebiotic synthesis of biopolymers on inorganic templates," *Naturwiss,* 1973, 60.

Kerkut, G. A. *The Implications of Evolution,* London: Pergamon Press, 1960.

Kitts, D. B. "Paleontology and Evolution Reconsidered," *Paleobiology,* 1977, 3.

Kohler, J., S. Schafer-Preuss, D. Buttgereit, "Related enhancers in the intron of the beta1 tubulin gene of Drosophila melanogaster are essential for maternal and CNS-specific expression during embryogenesis." *Nucleic Acids Res* 1996, 24.

Kuhn, Thomas. *The Structure of Scientific Revolutions.* Chicago: University of Chicago Press, 1962.

Lemaitre, Georges. "Un univers homogène de masse constante et de rayon croissant, rendant compte de la vitesse radiale des nébuleuses extragalactiques," *Annales de la Société scientifique de Bruxelles* 1927, 47.

Levin, R. "Bones of Mammals, Ancestors Fleshed Out." *Science,* Vol. 212, 26 June 1981.

Lingenfelter, Richard E. "Production of C-14 by Cosmic 8 Ray Neutrons." *Reviews of Geophysics,* 1:51, February, 1963.

Macbeth, Norman. *Darwin Retried: An Appeal to Reason.* Boston: Gambit, 1971.

Makalowski, W. "Not Junk After All" *Science,* 23 May 2003, Vol. 300. No. 5623.

Matthews, L. Harrison. "Introduction" to *The Origin of Species* by Charles Darwin, 1971 edition. London: J. M. Dent and Sons, 1971.

Maeda S., and G. Mogi. "Functional Morphology of Tonsillar Crypts in Recurrent Tonsillitis," *Acta Otolaryngo (Stockh) Suppl*, 1984, 416.

Mayda, Arslan. "İşe yaramaz zannedilen kuyruk sokumu." *Sızıntı*, 1997, No. 227, Izmir.

McCann, Lester J. *Blowing the Whistle on Darwinism*, Lester J. McCann, 1986.

McMullen, T. E. "Problems with chemical origins of life theories," Excerpts from his lectures between April 16, 1993 and April 3, 1995 at South Carolina University. (http://www2.gasou.edu/facstaff/etmcmull/CHEM.htm).

McNamara, Ken. "Embryos and Evolution," *New Scientist*, October 16, 1999.

Meyer, Stephen C., Scott Minnich, Jonathan Moneymaker, Paul A. Nelson, and Ralph Seelke, *Explore Evolution: The Arguments for and Against Neo-Darwinism*. Melbourne: Hill House Publishers, c/o O'Brien & Partners, 2007.

Miller, Stanley L. "A Production of Amino Acids Under Possible Primitive Earth Conditions," *Science*, Vol. 117, May 15, 1953. No: 3046.

———. "Current Status of the Prebiotic Synthesis of Small Molecules," *Molecular Evolution of Life*, 1986.

Miller, Stanley L., and H. C. Urey, "Organic Compound Synthesis on the Primitive Earth," *Science*, 1959, 130.

Milton, Richard. *Shattering the Myths of Darwinism*. Vermont: Park Street Press, 1997.

"Missing, Believed Nonexistent," *The Guardian Weekly*, November 26, 1978, vol. 119, no 22, in Denton 1988.

Mora, P. T. "The Folly of Probability" in *The Origins of Prebiological Systems and Their Molecular Matrices*, edited by Sidney W. Fox. New York: Academic Press, 1965.

Morowitz, Harold J. *Energy Flow in Biology*. New York: Academic Press, 1968.

Mottran, V. H. "In the Organ Corporation," *Liner*, April 22, 1948.

Nelson, Laura. "First chimp chromosome creates puzzles," *Nature Science Update*, May 27, 2004.

Nouy, Pierre Lecompte du. *Human Destiny*. London: Longmans Gren and Co., 1947. First Ed.

Nursi, Bediüzzaman Said. (1976). *Lem'alar*. Risale-i Nur Külliyatından. Istanbul: Sözler Yayınevi, 1976.

Officer, Charles B., and Drake, Charles L. "The Cretaceous-Tertiary Transition," *Science* 1983, 219.

"Old Bird," *Discover*. March 1997.

"On Campus," Alleged skullduggery, Random Samples. *Science*, Vol 305, Issue 5688, August 27, 2004.

Oparin, Aleksandr Ivanovich. *Life: Its Nature, Origin and Development*. London: Oliver&Boyd 1961. Translated from Russian by Ann Synge.

Oppenheimer, J. M. "Haeckel's variations on Darwin," *Biological Metaphor and Cladistic Classification: An Interdisciplinary Perspective*. Edited by H. M. Hoenigswald and L. F. Wiener. Philadelphia: University of Pennsylvania Press, 1987.

Paterson, Tony. "Neanderthal Man never walked in northern Europe." www.telegraph.co.uk/news/main.jhtml?xml=/news/2004/08/22/wnean22.xml. August 22, 2004.

Patterson, C. *Harper's*, February 1984.

Pennisi, Elizabeth. *Science News*, December 10, 1994.

———. "Haeckel's Embryos: Fraud Rediscovered," *Science* Vol. 277, No. 5331, September 5, 1997.

Penzias, A. A., and R. W. Wilson, "A Measurement of Excess Antenna Temperature at 4080 Mc/s," *Astrophysics Journal* 1965, 142.

Popper, Karl Raimund. *Unended Quest: An Intellectual Autobiography*. Illinois: Open Court, 1976. The Library of Living Philosophers, Vol. 1.

———. "Darwinism as a metaphysical research programme." *Methodology and Science*, 1976, 9.

Raup, David. "Conflicts between Darwin and Paleontology," *Field Museum of Natural History Bulletin*, vol. 50. No. 1, 1979.

Raup, David, and Sepkoski, Jack. "Periodicity of Extinctions in the Geologic Past." *Proceedings of the National Academy of Science*, 1984, 81.

Renauld, H., and S. M. Gasser, "Heterochromatin: a meiotic matchmaker," *Trends in Cell Biology 7*, May 1997.

Rensberger, Boyce. *Houston Chronicle*, November 5, 1980.

Richardson, Michael K., J. Hanken, M. L. Gooneratne et al. "There is no highly conserved embryonic stage in the vertebrates, implications for current theories of evolution and development," *Anatomy and Embryology*, 1997, 196.

Richardson, Michael K. et al. "Haeckel, Embryos, and Evolution," *Science*, May 15, 1998, 280.

Richardson, Michael K., and Gerhard Keuck, "A question of intent: when is a 'schematic' illustration a fraud?" *Nature* 2001, 410:144.

Richardson, Michael K., and Gerhard Keuck, "Haeckel's ABC of evolution and development," *Biological Reviews of the Cambridge Philosophical Society* 2002, 77.

Richardson, Michael K. "Haeckel's Embryos, Continued," *Science*, 1998, 281.

Ridley, Mark. "Who doubts evolution?" *New Scientist*, 25 June 1981, Vol. 90.

Rifkin, Jeremy. *Algeny: A New Word, A New World*. Penguin: 1984.

Rutimeyer, Ludwig. "Rezension zu Haeckel, Ernst, Naturliche Schöpfungsgeschichte." Berlin: 1868, *Archiv für Anthropologie* 3.

Sagan, Carl. "Life," *Encyclopedia Britannica*. (New York: Encyclopedia Britannica, 1997), 22.

Salisbury, Frank B. "Natural Selection and the Complexity of the Gene," *Nature*, 1969, 224.

Sandell, L. L., and V. A. Zakian. "Loss of a yeast telomere: arrest, recovery, and chromosome loss" *Cell* 1993, 75 (4).

Schulz, Matthias. "Die Regeln Mache Ich," *Der Spiegel*, August 16, 2004.

Sepkoski, John, Jr., "Rates of speciation in the fossil record." *Philosophical Transactions of the Royal Society of London B: Biological Sciences*, 353 (1366).

Simpson, George Gaylord. *Horses*. Oxford University Press, 1961.

———. *The Major Features of Evolution*. New York: Columbia University Press, 1961.

———. *The Meaning of Evolution*. Revised Edition. New Haven, Connecticut: Yale University Press, 1967.

———. *Life Before Man*. New York: Time-Life Books, 1972.

Simpson, George Gaylord, and W. Beck, *An Introduction to Biology*, New York: Harcourt Brace and World, 1965.

Spencer, Herbert. *First Principles of a New System of Philosophy*. New York, Appleton, 1872. Two volumes.

Stahl, B. J. *Vertebrata History. Problems in Evolution*. New York: McGraw-Hill, 1985.

Stanley, Steven M. "Mass Extinctions in the Ocean." *Scientific American*. No: 6, June 1984.

Staune, Jean. "L'évolution condamne Darwin." The interview with *Jean Dorst*. *Figaro Magazine*. October 26, 1991.

Suess, Hans E. "Secular Variations in the Cosmic-Ray produced Carbon-14 in the Atmosphere and Their Interpretations." *Journal of Geophysical Research*, 70:5947, December 1, 1965.

Swinton, W. E. "The Origin of Birds" in *Biology and Comparative Physiology of Birds*, A. J. Marshall (edited by) New York: Academic Press, vol. 1.

Switzer, V. R. "Radioactive Dating and Low-level Counting," *Science*, 157:726, August 11, 1967.

Taylor, G. R. *The Great Evolution Mystery*. New York: Harper & Row, 1983.

Thomson, John Arthur, and Geddes, Patrick. *Life: Outlines of General Biology*. London: Williams & Norgate 1931. Vol. II.

Thomson, Keith Stewart. "Ontogeny and Phylogeny Recapitulated,"*American Scientist*, Vol. 776, May–June 1988.

Thompson, W. R. "Introduction" to *The Origin of Species* by Charles Darwin, 1956 edition. New York: E. P. Dutton, 1956.

Ting, S. J. "A binary model of repetitive DNA sequence in Caenorhabditis elegans." *DNA Cell Biol*. 1995, 14.

Vandendries, E. R., D. Johnson, R. Reinke, "Orthodenticle is required for photoreceptor cell development in the Drosophila eye." *Dev Biol* 1996, 173.

Vogel, Gretchen. "Objection 2: Why Sequence the Junk?" *Science*, February 16, 2001.

Waddington, Conrad Hal. *The Strategy of the Genes*. London: Allen-Unwin, 1957.

Ward, Peter, and Brownlee, Donald. *Rare Earth*. New York: Copernicus, 2000.

Watanabe, H., and E. Fujiyama, et. al. "DNA sequence and comparative analysis of chimpanzee chromosome 22," *Nature*, 2004, 429.

Weiner, W. S., K. P. Oakley, W. E. Le Gros Clark, "The Solution of the Piltdown Problem," *Bulletin of the British Museum (Natural History) Geology Series*, 1953, Vol. 2, No. 3.

Wells, Jonathan. *Icons of Evolution: Science or Myth? Why Much of What We Teach about Evolution Is Wrong*. Washington DC: Regnery Press, 2000.

Westoll, Thomas Stanley. Proceedings from the British Association Meeting at Edinburgh, August 10, 1951.

Wilder-Smith, A. E. *The Natural Sciences Know Nothing of Evolution*. California: Master Books, 1981.

Wood, B., and A. Brooks. "We are what we ate," *Nature*, 1999, Vol. 400, no: 6741, 15 July 1999.

Wysong, R. L. *The Creation-Evolution Controversy*. East Lansing, MI: Inquiry Press, 1976.

Yockey, Hubert P. "A Calculation of the Probability of Spontaneous Biogenesis by Information Theory," *Journal of Theoretical Biology*, 1977, 67.

Yockey, Hubert P. *Information Theory and Molecular Biology*. Cambridge University Press, 1992.

Zuckerkandl, E. "Neutral and Nonneutral Mutations: The Creative Mix-Evolution of Complexity in Gene Interaction Systems," *Journal of Molecular Evolution*, 1997, 44.

INDEX